Gram Stain:
Looking Beyond Bacteria
TO FIND FUNGI IN GRAM STAINED SMEAR

A LABORATORY GUIDE FOR MEDICAL MICROBIOLOGY

….. a textbook as timeless as the Gram stain procedure first discovered by Christian Jacob Hans Gram in 1883 …

Explore this textbook on Gram stain and learn simple techniques to identify the most important clinical fungal isolates recovered from clinical specimens.

BY: SUBHASH K. MOHAN, GAMS (INDIA), MLT, ART (CANADA)

SENIOR AND TEACHING MEDICAL MYCOLOGY TECHNOLOGIST
UNIVERSITY HEALTH NETWORK — TORONTO MEDICAL LABORATORY AND MOUNT
SINAI HOSPITAL, TORONTO, ONTARIO, CANADA

AuthorHouse™
1663 Liberty Drive
Bloomington, IN 47403
www.authorhouse.com
Phone: 1 (833) 262-8899

*Because of the dynamic nature of the Internet, any web addresses or links contained in this book may have changed
since publication and may no longer be valid. The views expressed in this work are solely those of the author and do not
necessarily reflect the views of the publisher, and the publisher hereby disclaims any responsibility for them.*

*Any people depicted in stock imagery provided by Getty Images are models,
and such images are being used for illustrative purposes only.
Certain stock imagery © Getty Images.*

This book is printed on acid-free paper.

ISBN: 9781438960289 (sc)

Library of Congress Control Number: 2009902860

Print information available on the last page.

Published by AuthorHouse 09/30/2020

author HOUSE®

This work is dedicated to my brother, **Sudarshan Kumar Mohan,** who was my mentor and a role model.

THANK-YOU NOTE

The time and dedication required to complete this technical laboratory guide has been fully supported by my wife **Aruna** and our only daughter **Monica,** who spent many weekends and uncounted evenings in my absence.

And special thanks to **Dr. Heather J. Adam** and **Dr. Patrick Tang** for providing helpful ideas, suggestions and unconditional support for making my dream project a reality.

Acknowledgements

I sincerely acknowledge the following individuals for their encouragement that led to developing this technical manual for medical microbiology.

Dr. Iivi Campbell, Dr. Donald E. Low, Dr. Tony Mazzulli, Dr. Kelly S. MacDonald, Dr. Allison McGeer, Dr. Susan M. Poutanen, Dr. Leslie Spence, Dr. James Brunton, Dr. Ian E. Crandall, Dr. Ralph G. Albright, Dr. Suresh Syal, Dr. Richard Summerbell, Professor Lynne Sigler, Dr. Susan E. Richardson, Dr. H. Roslyn Devlin, Dr. Anne M. Phillips, Dr. Mel Krajden, Dr. Christine H. Lee, Dr. Andrew E. Simor, Dr. Glen D. Roberts, Dr. Elmer W. Koneman, Dr. Mabel Rodrigues, Dr. Lakshmi Gannavarapu, Dr. Moira Grant, Dr. Kanchana Manickam, Dr. Jeff Fuller, Dr. Deepali Kumar, Dr. Atul Humar, Dr. Pratima Deb, Dr. Hani Dick, Mr. William A. Marshall, Mr. Purshotam Bhatnagar, Mr. Martin Skulnick, Ms. Katherine Wong, Ms. Ann Gray, Ms. Audrey Jarrett, Ms. Linda Zuchowski, Ms. Beulah Mustachi, Ms. Amy Costales, Ms. Yin-Ping Tse and Ms. Suzie Ho.

CONTENTS

By: Subhash K. Mohan, GAMS (India), MLT, ART (Canada) i

Introduction

Chapter 1

Chapter 2

Chapter 3

Chapter 10

Chapter 11

Chapter 12

Chapter 13

Chapter 14

Chapter 15

Chapter 16

Chapter 17

Chapter 18

Chapter 19

Chapter 20 The quizzes

INTRODUCTION

Although there are several medical mycology textbooks available on the market, each containing a chapter on direct microscopy, none has addressed the use of the Gram stain for finding fungi. This book will be the first to advocate the use of the Gram stain as an additional method in clinical microbiology and mycology laboratories for detecting fungi. The Gram stain is often neglected or considered suboptimal for this purpose. However, with the trained eye, much additional information can be gleaned from the Gram stain. This book will describe many scenarios where the Gram stain has been useful for the laboratory diagnosis of fungal infection. When a fungal infection is present but not suspected clinically, the Gram stain may be the only clue to the true cause of the infection. Although all medical-laboratory technologists will recognize bacteria and yeast in Gram-stained smears, most will miss the presence of other fungal elements. Identifying fungal elements in the Gram stain is another important indication that a positive fungal culture is clinically relevant instead of an environmental contaminant. This book is based on the innovations and experience of the author and hopes to introduce an additional tool to the very subjective field of medical mycology.

The direct detection of fungi in clinical specimens is a crucial part of medical mycology. Fungi grow at variable rates and often slowly. Therefore, the identification of fungi in a smear is crucial in providing a timely presumptive diagnosis and may guide earlier initiation of appropriate antifungal therapy. Although there are better methods than the Gram stain for visualization of fungi, these methods are only performed if there is clinical suspicion for fungal disease. Clinicians often send specimens for bacterial culture, but requests for fungal culture are not made unless there is strong suspicion of fungal infection. In such cases, the Gram stain is the only technique available in the clinical microbiology laboratory for direct examination of these specimens. Most medical-laboratory technologists are not trained to recognize fungal elements, other than yeasts, in Gram-stained smears. This book will guide the reader in the recognition and identification of fungal elements in Gram-stained smears, especially when they are distorted or remain unstained and undetectable.

The Gram stain should not be overlooked during direct examination of the clinical specimens in searching for bacteria. Gram stains can also turn into a method for fungal diagnosis in addition to bacteria. While concentrating on a microscopic object, gathering enough information, clue, and effort, the average medical-laboratory technologist should be able to identify fungal elements in Gram-stained smears. When fungal cultures have not been specifically requested, the identification of fungi in the Gram stain will allow the technologist to offer additional laboratory information that may have a significant impact on the clinical management of the patient.

There are times during examining of the direct Gram smear when cells other than bacteria are observed and the smear reader questions whether you would detect a round or oblong cellular structure in a Gram stain and be able to classify it as debris or yeast. Astute Gram stain skills have the potential to immediately and positively impact the quality of patient care. This new textbook focuses on the detection and classification of fungal elements in Gram stains. Newly developed flowcharts have been designed to guide you in the detection of fungal elements. Many clues and key details regarding structural characteristics are described that are essential to distinguish fungal elements. The advantages and disadvantages of different staining procedures and the utility of routine bacteriology staining procedures, such as the Gram stain, in the interpretation of microscopic structures are discussed. Proficiency in the detection and identification of fungal elements is paramount to the successful treatment of serious medical conditions, especially when a fungal cause has not been suspected.

The detection of fungal elements in direct smears enables a more direct correlation with the clinical presentation than fungal cultures grown weeks later. The ability to rapidly detect fungal elements in a clinical specimen enables prompt antifungal therapy when needed without the associated concern of unnecessary treatment of a potential contaminant. Throughout the years, the author has accumulated many scenarios in which fungal elements were not detected on the original Gram stain evaluation but were found to be positive upon review once the cultures grew fungus. Many of these scenarios and possible explanations for the misread smears are described in the text.

There are several medical mycology textbooks that contain a chapter on direct microscopy. However, this textbook is the first of its kind as it discusses the simple Gram stain procedure as a valuable tool for the detection of fungal elements. Key questions addressed include: What difficulties do medical technologists encounter during direct examination of the clinical specimens? How can the mycology laboratory possibly differentiate pathogenic fungus from environmental contamination?

The clinical microbiology lab receives many specimens requesting bacteriology culture and sensitivity. A medical technologist examining the Gram smears of these specimens generally searches for bacteria and yeasts but frequently missed the presence of fungal elements and other organisms. Fungal elements are often interpreted as tissue structures or debris. In addition to the flowcharts of fungal element interpretation in Gram stains, this textbook includes a protocol for staining over the top of a Gram stain to permit greater visualization of fungal elements.

Direct detection of fungal elements in clinical specimens using routine Gram stain procedures improves the microbiology laboratory's ability to provide presumptive diagnosis and guide patient care, since fungal causes are often unsuspected. Based on the microscopic observations in the clinical specimens, appropriate media can be selected to improve the yield in the laboratory and may also alert staff for special precautions. Although medical technologists are not currently taught to examine Gram-stained smears for fungal elements, it is essential to the patient care that they begin to do so.

CHAPTER 1

1.1 HISTORY OF THE GRAM STAIN

The utility of Gram stains for the differentiation of bacteria was discovered by a Danish doctor Christian Jacob Hans Gram in Germany in 1883.[1, 13, 14, 17, 18, 39] The ability to differentiate bacteria by this method was an incidental discovery made by Gram as part of his attempts to stain kidney tissue to distinguish urine casts. Several modifications have been made to the original Gram stain protocol in order to make the procedure suitable for general lab use; however, the basic principle remains unchanged. The Gram stain was devised by Hans Gram to stain, detect, and categorize bacteria in two distinct groups as Gram positive and Gram negative.

The basis of the Gram stain reaction is related to two main things: the thickness of the cell wall and the pH of the cellular material within cytoplasmic membrane. The target of the Gram reaction is not the cell wall but the interior of the cell, where positive ions of colored basic stain bind negatively charged cellular material in acidic environments.[1, 13, 14, 17, 18, 39] Primary stain is bound inside the cellular components and is unable to be removed from the thick-walled Gram positive bacteria due to the larger-sized complex formed between crystal violet (CV) and iodine. Decolorizer use (causing large pores in the cell wall, unable to hold CV-iodine complex) would eliminate all the CV from the Gram negative bacteria and would stain pink when the counterstain safranin is applied.

Hans Gram, born on September 13, 1853, in Copenhagen, first studied botany and later completed medicine. He worked in several hospitals in Copenhagen and received gold medal from the university for his hematological studies and continued to work until he enrolled in the microbiology course run by Carl Salomonsen. Hans Gram was referred by Salomonsen and joined Dr. Carl Friedlander's department of pathology in Berlin on October 22, 1883. Within days, Gram got interested in staining kidney parenchyma and tubular casts differentially without elaborating its reason. Gentian violet in aniline water was introduced by Ehrlich, which stained nuclei, fibrin, bacteria, and tubular casts. However, the staining intensity would not allow bacteria to be visualized differentially.

Gram started to look for a solution using gentian violet to stain bacteria but not the tissue so that the microorganisms could be picked up under the microscope. Gram used alcohol after gentian violet, which removed the stain from the tissue, including tubular casts. Gram was a little disappointed that he was unable to separate tissue cells from casts since the staining dye was removed by decolorizer; but he noticed that bacteria remained unaffected by the use of alcohol. Gram used bismark brown as a counterstain. This work led to the discovery of the Gram stain within two weeks after Gram joined Friedlander's laboratory, and he published his method in 1884. Gram's letters written to Salomonsen reveal that Gram used iodine in his method to demonstrate purple color retained by pneumococcus causing lobar pneumonia. However, Friedlander was not impressed by this procedure since he was working on capsulated bacteria that did not show Gram positive reaction. Both Gram and Friedlander were working on two different bacteria; one worked on Gram positive coccus and the other was playing with Gram negative bacillus. It was later confirmed that the organism Friedlander was working on was actually known as *Klebsiella pneumonia* (formerly called Friedlander's bacillus). Both organisms caused pneumonia in different patients, and both Hans Gram and Friedlander worked on different set of clinical cases as well. In March/April 1884, Friedlander fully acknowledged Gram's achievement about the Gram stain method before he went on leave due to tuberculosis and died in 1887 at age 40.[17, 18]

Hans Gram, after successfully decolorizing bacteria, using an iodine decolorization method as mentioned in one of his four letters dated January 2, 1884, that was published 10 weeks later. Hans Gram did not continue his work on the Gram stain technique. His fifth letter from Berlin, on January 22, 1884, expresses his concern about Friedlander's capsulated cocci, and he was reluctant to call these cocci bacilli although they were decolorized by his method and stained pink. Friedlander was not pleased with the iodine method and asked Hans Gram to withhold his paper from publication. Friedlander published Gram's iodine method as mentioned in Hans Gram's sixth and the last letter from Berlin in January 28, 1884, announcing that his iodine method is a valid technique to differentiate bacteria.[17, 18] In other words, two distinct categories of bacteria were created that helped the taxonomy of Gram positive and Gram negative bacteria. The credit goes to Hans Gram and Friedlander for providing a meaningful interpretation about bacteria, dividing them into two groups and helping clinicians to choose appropriate antibiotics in order to treat patients suffering from bacterial infections.

Hans Gram left Berlin on March 20, 1884, and traveled to Strasbourg where he decided to work in a pharmacology laboratory and wrote three letters from Strasbourg to Salomonsen. In his first letter, on June 3, 1884, Hans Gram referred to Friedlander's investigations about a single case of pneumonia that Friedlander considered to be cocci overly decolorized by iodine treatment, which in reality was Gram-negative bacilli colored in pink; back then this was known as Friedlander's bacillus and was later named *Klebsiella pneumoniae*, which is actually a Gram negative bacillus. In Gram's letter to Salomonsen, he describes that Friedlander now shared his belief to higher degree. The confusion between the two doctors (Hans Gram and Friedlander) was mounted by the fact that both worked on different bugs unknown to both at the time during their staining the organisms separately. Both experts tried to convince the other about the facts they gathered from Gram stain reactions without knowing that Gram positive and Gram negative organisms have caused similar pneumonic symptoms in different patients. The problem was resolved in 1886 when Weichselbaum found that 94 of 129 patients having pneumonia were infected by *Diplococcus pneumoniae*, now known as *Streptococcus pneumoniae* (Gram positive cocci), and nine lungs of the patients were infected by Friedlander's bacillus (*Bacillus pneumoniae*), now known as *Klebsiella pneumoniae* (Gram negative bacilli).[1, 17, 18]

On September 15, 1938 (two days after his 85th birthday), Hans Gram published his article on serum treatment for type-III pneumococcal pneumonia. Two months later Hans Gram died. Scientists have made several accidental discoveries by observing something unexpected that had not been known, planned, or seen. Hans Gram's staining technique is one of them.[1, 17, 18]

The history of Gram stain reveals that Hans Gram was successful in discovering that Gram stains differentiate bacteria into Gram positive as purple and Gram negative as pink. However, the morphological grouping of bacteria may have arrived at a later date. Gram's reaction was greatly influenced by the cell wall, since the cell wall acting directly to prevent decolorization was considered a significant factor over the binding of the basic dyes to the interior cellular material of the bacterial cell. This has been discussed during the "Twenty-Ninth Annual Meeting of the Society of American Bacteriologists, School of Medicine and Dentistry" at the University of Rochester, Rochester, New York, on December 28, 29, and 30, 1927—that morphology should be the sole basis for differentiation of genera (grouping of microorganisms). This helped a great deal by grouping organisms according to the specified morphological features followed by the Gram reactions to make organisms fit into systematic organized fashion. Therefore the morphology as well as the Gram reaction became the basis of creating several groups of microorganisms stained as Gram positive or Gram negative by Hans Gram's iodine technique, which has been modified over the years to make it more useful and reproducible, except that the basic theory of the Gram stain remains unchanged.[1, 17, 18]

The basic dyes bind with carboxyl groups in the bacterial cell components; therefore the staining reaction occurs at a specific chemical site, suggesting the function of mass action laws in the staining of bacterial cells.[1, 3, 4, 11] Several factors influenced the reaction of Gram stain; therefore it required careful control of all the steps in the Gram stain to prevent false results causing confusion and hindering correct interpretation of the Gram reaction. Timing every step in the procedure in the Gram stain method became the hallmark for staining bacteria in clinical specimens. However, later studies found that the Gram stain procedure may be adversely affected by one basic dye replacing another, low pH, and low concentration of basic dyes.[1, 2, 3, 4, 11, 39] As a result, the Gram reaction may produce suboptimal staining pattern, causing difficulty in interpretation.

However, others claim that the Gram stainability is a function of the cell wall and not related to the chemistry of the cell constituents. There are so many thoughts about the Gram stain reaction; some claim that the staining reaction is the basic dye binding to the acidic environment of the cell, while others claim that the cell wall thickness is responsible for retaining the dye behind the cell wall. Technically speaking, the acidic pH in the interior of the bacterial cell is responsible for binding the positively charged ions of the basic dyes, although the degree of unbound and bound dye would be affected by the type of decolorizer used as well as the duration of the decolorizer applied. Thick cell walls are a mechanical hindrance for retaining the large complex of CV-iodine unable to escape, but it is removed from the cell having large pores in the cell wall. The cell wall damage due to decolorizer results in loss of cell wall integrity, causing problems in bacteriology where Gram positive organisms began to appear Gram negative or variable (i.e., visualization of certain bacteria in Gram-stained smears becomes uninterruptible).

For practical purposes it does not matter what caused the CV to be held back behind the cell wall of purple-colored Gram positive bacteria. The line drawn between the two categories of bacteria is too important to separate one group of bacteria from the other. Pink-stained Gram negative bacteria could not hold CV-iodine complex behind the wall. The separation of Gram positive and Gram negative bacteria is dominated by the color contrast developed by the Gram reaction influenced by many physical and chemical variables. Gram stain divides bacterial groups into two distinct entities. All other organisms staining purple or pink do not become valid candidates for the Gram stain reaction, since it does not divide other organisms into different groups nor help to alter the therapy based on Gram reaction that is the most important information provided by the Gram reaction for bacteria only.

The peptidoglycan layer in the cell wall of Gram positive bacteria is 10–15 times thicker than the cell wall of Gram negative bacteria.[1, 9, 17, 30, 45] The cell wall components in the fungal cell wall are different and even thicker as compared to the cell wall of bacteria.[3, 16, 17] The thickness of the cell wall plays a major role in the retention of basic dyes regardless of the cellular composition to which CV may bind. The ultimate retention of the CV-iodine is due to the physical properties of the cell wall unrelated to its chemical composition.

Scientists have yet to agree on the exact definition of the Gram stain. Should we call all organisms staining purple by the Gram stain Gram positive without worrying about thick-cell-walled organisms retaining CV? Similarly, should we call all organisms staining pink Gram negative without paying attention to the cell wall thickness that allowed CV-iodine complex to leak from the Gram positive cell wall that has been damaged extensively due to the prolonged application by decolorizer? The Gram stain reaction is far more meaningful for categorizing bacteria into two distinct groups but has no value for other organisms, such as fungi staining purple or pink by the Gram stain procedure. However, the Gram stain is extremely useful for the detection of organisms other than bacteria regardless of the color they display. Binding of the CV in the interior of the bacterial cell is the hallmark for the Gram positive reaction influenced by the thickness of cell wall. However, the thick cell wall of fungi not only plays a role in controlling the chemistry by restricting the CV-iodine complex from escaping and keeping it behind the boundary of the cell wall displaying staining reaction as Gram positive. Many times fungal cell walls also disallow the primary basic dye CV to diffuse inside the cell, turning the organism Gram negative only after the application of the decolorizer digs enough holes in the cell wall for safranin to get in and display pink coloration. Those fungal elements not allowing any basic dye a passage to the interior of the cell would appear unstained. Such fungi remain unaffected by the decolorizer and as a result would not allow entry of any basic dye (CV and safranin). Similarly, some known Gram positive bacteria (archaeobacterium) do not allow CV to diffuse through the thick cell wall and stain Gram negative. However, some species within the same group having cell wall thickness similar to Bacillaceae retain Gram stain (CV), indicating that the Gram stain reaction is not absolutely dependent on cell wall thickness.[1, 21, 39]

1.2 FOCUS ON THE GRAM STAIN

Medical-laboratory technologists must keep thinking and observing before, during, and after the Gram stain procedure has been performed on clinical specimens to be examined microscopically. It has frequently been seen that the Gram stain procedure being so simple does not always result in optimum performance. The medical-laboratory technologists often come across over- and under-decolorizing problems and suspect "decolorizer" as the troublemaker. However, there are several other things that play roles in turning Gram-stained smear into a problematic issue. The technologists must consider the Gram stain method credibly sophisticated rather than assuming it to be a simple and easy test procedure. It must also be remembered that the simplicity and the experience must not be allowed to influence the technologist to skip steps during staining a smear using the Gram stain procedure only because it is so simple and easy to perform. The following suggestions would help technologists to maximize the clarity and quality of the Gram stain technique for the detection of microorganisms in the clinical specimens.

1.3 WHAT SHOULD YOU DO BEFORE STAINING A GRAM SMEAR?

- Check the expiration date of the regents
- Appropriate concentration of crystal violet is essential (~ 2%)
- Check iodine concentration (0.33% to 1%); if it appears lighter, replace it
- Make a thin smear and air-dry the preparation before fixing
- Gently heat-fix the smear (or fix with methanol); avoid overheating
- Do not forget to include QC smears for staining

1.4 WHAT SHOULD YOU DO DURING STAINING A GRAM SMEAR?

- Avoid excessive washing in between staining steps
- Do not extend time of the reagents applied to the smear beyond specified mark

•Do not prolong application of decolorizer on a thin smear

•Make sure enough time for decolorizer is allowed for a thick smear

•Do not leave counterstain on for too long at the end

1.5 What should you do after staining a Gram smear?

•Let smear stand to allow water to roll down to dry

•Do not dry smears on electrical hot metal plate or other heating devices

•Do not blot smear in between tissue paper

•Observe stained smear visually for quality before reading

Chapter 2

2.1 Fungi[17, 29, 30, 45]

Over one million fungal species exist in nature. However, less than a few hundred may be clinically important. About 50 fungal species may be able to cause infections in normal and healthy individuals. More and more fungi are now becoming opportunistic pathogens in immunocompromised and debilitated hosts. The natural habitat of fungi is soil, plants, wood, compost, and decomposing organic material.[17, 30] Fungi may not be able to dig deep inside human tissue to cause infections. Fungi do not share human metabolism. Fungi require preformed organic carbon compounds for nutrition,[29, 30, 45] giving the author a reason to believe that fungi may have difficulty breaking down complex organic compounds. Therefore, fungi have three main barriers to overcome before getting established in the human tissue, such as temperature, immunity and human tissue as a source for nutrition. Many fungi are naturally thermo-tolerant. However, thermo-tolerant fungi appear to have low virulence as well as other factors discouraging their establishment in the human host to cause infection. Similarly, certain fungi are site specific and remain in superficial areas equivalent to natural and favorable environments away from immune response, temperature, and complex human tissue. In order for fungi to gain access and feed on human tissue, both (fungus and the human host) have to change to accommodate each other. Humans change by becoming immunocompromised (by choice or chance), while fungi bring about structural changes in them as "best fit" in the host's living tissue environment.

Patients having fungal infections are examined by the clinician, and the clinical specimens are collected for fungal analysis. Fungi present in the clinical specimens are processed to recover from the corresponding culture. At the same time, the single most important test to detect fungi in the patient's specimen such as direct microscopy is carried out pending culture confirmation and the full identification of the fungus involved in the clinical setting.

2.2 Direct Microscopy

Direct microscopy is the single most useful procedure in clinical microbiology/ mycology. Frequently, a provisional diagnosis may be made based on the fungal elements seen in direct smear. Microscopic examination of the specimen showing fungi allows the selection of appropriate media for culture as well as to alert for special precautions if potential pathogenic fungus of a dimorphic nature is seen. Direct microscopic examination of the clinical specimen using a Gram stain remains rapid, easy, and the most cost-effective procedure in the clinical microbiology.

Differentiation of yeast and mold is important, since yeast may represent normal colonization.[9, 13, 14, 20, 32] Technologists reading Gram-stained smears in routine bacteriology *must think about microbiology* and not keep focusing on bacteria only. Sometimes Gram smears serve as excellent alternatives to fungal-stained smears such as potassium hydroxide (KOH) or calcofluor white (CW). Sometimes fungi missed or not seen in Gram smears give an indirect notion about the fact that an important fungal species may be involved in the clinical specimen. Therefore, reviewing Gram smears is beneficial as well as confirming the identity of the fungal structure that remains in doubt using CW. When fungi are observed in the Gram-stained smears, they are categorized based on their specific structures, shape, and size. If round and spherical structures are seen, separate them based on budding versus non-budding. If budding forms are seen, determine the attachment as narrow-based budding versus broad base. Distinguish cells as thin-walled, thick-walled, or double-walled. If filamentous forms are seen, observe the width of the hyphal fragment. If the width is <1μm consider it bacterial morphology. The width of filamentous organisms belonging to fungi is usually >1μm. If fungal hyphae are seen, separate true hyphae from pseudohyphae. True hyphae are further classified as septate and aseptate hyphae. Septate hyphae could be hyaline or pigmented. The author has devised flowcharts to simplify different categories of fungal elements seen in direct smear. Refer to **Flowcharts 1** and **2** to identify fungal elements based on the shape, size, and other identifiable structures.

FLOWCHART 1. MICROSCOPIC IDENTIFICATIONF OF FUNGAL ELEMENTS [30, 31]

Filamentous form
- Fungal hyphae
  - Hyaline — Septate / Aseptate / Pseudohyphae
  - Dark
- Bacillary form
  - Non-branching — Mycobacteria
  - Branching — Actinomycetes — Nocardia / Streptomyces / Other
  - Coccidioides / Rhinosporidium / Penicillium marneffei / Pneumocystis / Adiaconidia / Myospherule / Prototheca

Round and oval form
- Non-budding Cell (Coccidioides, Rhinosporidium, Penicillium marneffei, Pneumocystis, Adiaconidia, Myospherule, Prototheca)
- Budding Cell
  - Small
    - Narrow base — Torulopsis/ (Candida) / Histoplasma / Other
    - Unipolar — Malassezia
    - Variable — Sporothrix / Other
  - Large
    - Broad base — Blastomyces
    - Other — Histoplasma dubosoi / Paracoccidioides / Loboa loboi
    - Narrow base — Cryptococcus

FLOWCHART 2. DIFFERENTIATION OF ROUND AND OVAL CELLS SEEIN IN DIRECT SMEARS [10]

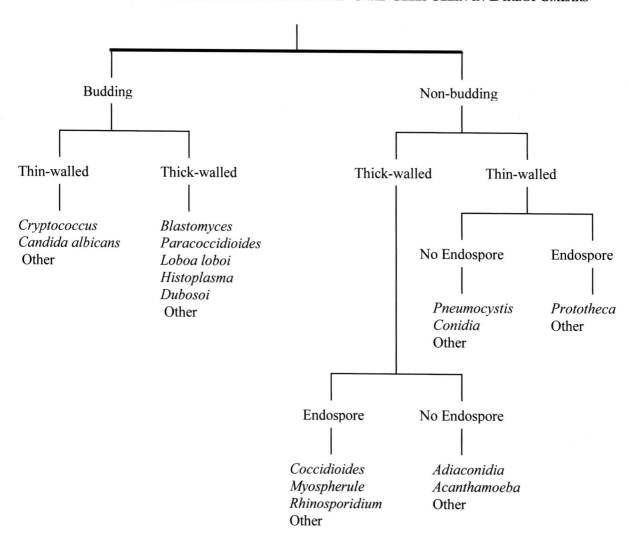

2.3 KEY FUNCTIONS OF DIRECT MICROSCOPY

-Rapid detection of microorganisms

-Provision of presumptive diagnosis

-Indication of clinical significance

-Determination of appropriate media

-Alert for safety precautions

2.4 OVERVIEW OF MICROBIOLOGICAL DIRECT STAINS

There are many kinds of stains utilized to detect and/or identify microorganisms in clinical specimens, as shown in **Figure** 1. These staining procedures can be divided into primary and specialized stains. The primary (routine) staining procedures are sufficient to detect fungi from most clinical specimens. Specialized stains are rarely required in the routine clinical microbiology except for confirmation or for the detection of certain fungi that fail to be identified by primary staining methods. The author's experience suggests that the primary routine staining procedures used in clinical microbiology are very useful and reproducible.

The Gram stain is the most commonly used procedure in the clinical microbiology laboratory. The utility of the Gram stain in clinical microbiology may be enhanced by the expansion of its reported detection ability to include yeast and filamentous fungi. Filamentous fungi may not always be readily recognized in the Gram stain, but increased experience and a little imagination make the detection of fungal elements possible. Other common stains such as Potassium Hydroxide (KOH), Giemsa, India ink, Calcofluor white (CW), Ziehl-Neelsen (ZN), and Modified Kinyoun (MK) are conducted as requested. KOH and CW are the most common stains used for the detection of fungi in clinical specimens.

Most laboratory use 10–20% KOH, an unstained but very sensitive method, to detect fungi in direct smears. KOH is a simple, rapid, and cost-effective procedure for the detection of fungal elements in the clinical specimens. Since KOH is a wet and unstained preparation, it requires experience for the correct interpretation of observed structures. Therefore, high rates of false negative and false positive reports may occur with less-experienced technologists. In order to improve clarity of fungal elements, a more sensitive fluorescent procedure such as CW is suggested. CW binds to the cell wall of fungi and fluoresces when exposed to long-wave UV light. CW may be added to the KOH preparation. KOH-CW preparations may be read using both the bright-field microscope and the UV microscope. The smear can be easily switched to bright field by shutting off the UV light and turning on the bright light filter to read the same smear and the same microscopic field. CW is a very useful method in mycology but has some limitations such as a lack of ability to detect yeast cells, including *Histoplasma capsulatum* and endospores of *Coccidioides immitis*. CW reagent binds to the surface of the cell wall, and as a result internal details of certain fungi are not fully resolved. Similarly, CW does not demonstrate pigmentation in the cell wall of dematiaceous fungi. The combination of KOH-CW provides the ease of examining the microscopic prep for fungal detection by using UV light; if needed to visualize further, the same prep may be examined by bright-field microscopy using white light. The majority of times simple staining procedures such as KOH, CW, and Gram procedures are sufficient for routine staining and finding fungi in the clinical specimens received by the microbiology laboratory.

Certain microbiology laboratories detect *Pneumocystis jiroveci* (formerly known as *Pneumocystis carinii*) from bronchoalveolar lavage (BAL) and induced sputum specimens as well as from other sites from immunocompromised patients using Fungi-Fluor (FF) procedure (Polysciences, Warrington, PA). FF is similar to CW and very useful for the detection of fungal elements and also allows the detection of

Pneumocystis jiroveci. FF procedure is a nonspecific, highly sensitive, simple, cost-effective, and comparable procedure to detect *Pneumocystis jiroveci*. The FF procedure to detect PCP in conjunction with GMS and/or IFA procedures serves as a back-up tool to avoid false positive and false negative tests results.

ZN and MK (both are kinyoun stains) stains are used to identify *Mycobacteria* and actinomycetes. Whenever *Nocardia* is suspected or identified in the direct Gram smear or requested specifically by the physician, MK and ZN are the two most important procedures to run in order to separate *Nocardia* from mycobacteria. ZN would be positive for acid fastness for Mycobacteria only, but not *Nocardia*. MK, on the other hand, would show partial acid fastness in *Nocardia* and also staining *Mycobacteria* as pink (red). The only difference in these two similar staining reagents is the decolorizer. Acid-alcohol decolorizer is used in ZN, and dilute sulphuric acid is used in MK. India Ink is the most frequently used stain to observe encapsulated *Cryptococcus* in CSF and BAL specimens. However, the low sensitivity makes this simple test less attractive, and it is replaced by latex agglutination test for Cryptococcal antigen detection.

Additional stains are usually carried out in histology departments when specifically requested by the physician or microbiologist. Stains such as Gomori Methenamine Silver (GMS) and Periodic Acid-Schiff (PAS) are specialized specific stains to detect fungi in the clinical specimens. GMS shows all fungi as dark against a green background. Smears prepared and stained in histology are excellent, with increased clarity. However, there are times when fungi stained by GMS are detected easily but are difficult-to-interpret structures seen under the microscope. Since GMS does not display cellular response, intracellular morphology of some important fungi would not be seen in GMS smear. Similarly, PAS stains fungi in optimum quality, including cellular response. However, the red-stained fungi and the background make detection a little difficult, especially in some small-sized yeast cells. Both GMS and PAS would not show pigment in the cell wall of fungi in the dematiaceous group.

Fontana Masson (FM) is used to demonstrate melanin pigment in the cell wall of dematiaceous fungi. Fungi in the dematiaceous group when seen in direct smear under the microscope may display pigmentation. However, many times only morphological clues of dark fungi help us to suspect their presence without the evidence of melanin pigment present in the cell wall of fungal elements seen under the microscope. CW does not demonstrate melanin pigment. However, suspected dark fungi may be observed under the bright-field microscopy, showing pigmented cell wall. Many times Gram-stained smears also show pigmented fungal elements. However, not all dark fungi would display melanin pigment in the cell wall. In these circumstances, it is important to determine melanin pigment in the cell wall of dematiaceous fungi using FM stain. Fungi containing melanin pigment in the fungal cell wall would appear brown to dark.

Mucicarmine (MC) is a specific stain for *Cryptococcus*, showing both the body and the capsule as a pink color. However, MC also stains *Blastomyces*. The differentiation of single non-budding cells of *Blastomyces* and *Cryptococcus* without a capsule is difficult with this method. The clue to separate one from the other is the double cell wall in *Blastomyces* and the thin wall observed in *Cryptococcus*.

Hematoxylin and Eosin (H&E) stain is another specialized stain done in the histology lab used to detect granuloma, areas of tissue (specimen) demonstrating cellular response to and around infectious organisms. H&E is also useful in the detection of some fungi and actinomycetes. Excellent structures of sulfur granular and actinomycotic mycetoma are visualized by this method.

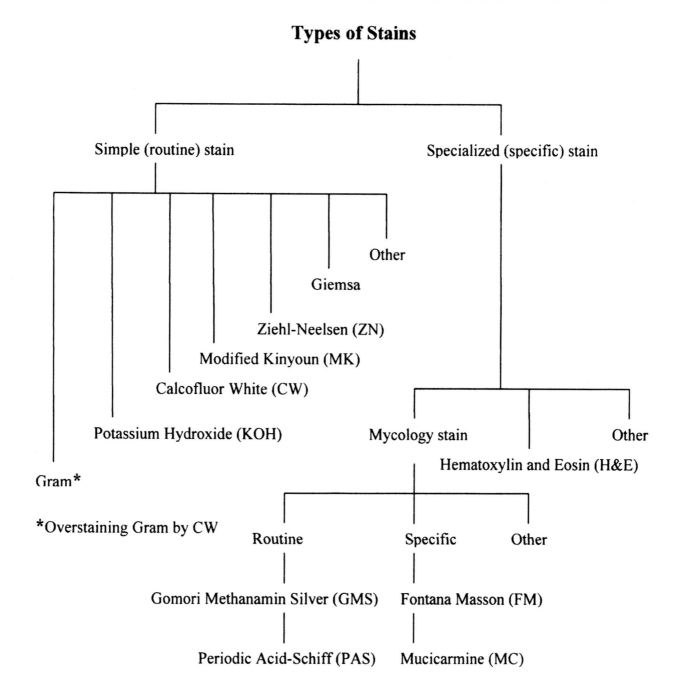

2.5 GRAM STAIN

The principle of Gram stain is related to the binding of CV (purple-colored) and safranin (pink-colored) dyes containing positive cations to the cellular components in the bacterial cell having negatively charged anions. The retaining of the colored dye (purple or pink) within the cell cytoplasm makes the distinction between two groups of bacteria; i.e., Gram positive (purple) and Gram negative (pink). This is the basic theory of Gram stain that plays a role in distinguishing bacteria based on color. However, this color-producing reaction is heavily dependent on two major partners in the process: *iodine* and *thick cell wall*. Iodine is essentially responsible for keeping CV behind the cell wall due to the formation of large molecules. The thick wall has a different role to play in different organisms, such as the thick cell wall in the Gram positive bacteria, which is tightly knitted with more protein molecules in it as compared to the thin cell wall in Gram negative bacteria containing more lipid than protein. As a result, the cell wall damage to the Gram negative bacteria caused by the decolorizer is vigorous and extensive, allowing all CV-iodine complexes to be washed away more easily after the application of decolorizer. The basic principle of the Gram reactions formed due to the binding of the Gram reagents and the nuclear material of the bacterial cell remains above all arguments. It is for this reason that all other cells (plant, fungi, human, and animal) should not be assumed as Gram positive or Gram negative based on the reagent they retain inside the cell that is purple or pink colored.

2.6 ARE FUNGI GRAM POSITIVE OR GRAM NEGATIVE?

Many textbooks indicate that fungi stain Gram positive, although there is no scientific explanation provided to support this observation. The rationale for fungi staining Gram positive may be inferred from the knowledge of cell structures. The cellular differences of bacterial, fungal, and human cells are listed in Table 1. Fungi have thick and rigid cell walls composed of chitin (similar to cellulose), glucan, chitosan, mannan, and glycoproteins. Animal cells have no cell wall; bacteria have no well-defined nucleus. Plant cells contain cellulose in the cell wall. Fungi and plant cells have a true nucleus.

TABLE 1. COMPARING THE CELL WALLS OF BACTERIAL, FUNGAL, AND HUMAN CELLS[1, 29, 33, 39, 45]

Cell Feature	Bacterial Cell	Fungal Cell	Human Cell
Nucleus	Circular, undefined	True nucleus	True nucleus
Cell Wall	Present	Present	Absent
pH	< 7	Unclear	Near 7
Cell Wall structure	**Gram positives:** peptidoglycan and lipoteichoic acid **Gram negatives:** lipopolysaccharide, phospholipid, and peptidoglycan	80%–90% polysaccharides and ~10% lipid and chitin	N/A

Fungal cell walls contain 80%–90% polysaccharides, and the rest is made up of protein and lipid. The skeletal material of the cell wall is spirally oriented fibrils of chitin (a polymer of N-acetyl-glucosamine), cellulose (a polymer of D-glucose), or sometimes other glucans. Protein is an important component; some of this may represent enzymes closely bound to the cell wall and therefore demonstrate biochemical activity. The walls of yeasts have distinct chemical properties composed of mannan-glucan complex,[7, 32, 39, 45] a complex structure composed of three major components: 1,3-beta-D-glucan, chitin (beta-1,4-N-acetylglucosamine), and mannoprotein (alpha-1,6-mannose). None of these elements are found in mammalian cells.[7, 29, 33]

The bacterial cell does not have a well-defined nucleus. It is negatively charged due to acidity. Basic dyes used in the Gram stain such as crystal violet and safranin are cations capable of binding to phosphate groups in the bacterial cytoplasm.[1, 2, 21] Initially, crystal violet and iodine bind nuclear material inside the bacterial cell. The cell wall of Gram positive bacteria contains more protein components than the cell wall of Gram negative bacteria. Similarly, the Gram negative cell wall contains more lipid components as compared to the cell wall of Gram positive bacteria.

The bacterial cell is an excellent candidate for the Gram stain procedure. All other cells such as human tissue, parasites, and fungi are technically stained by Gram reagents depending on what dye they retain without meaning anything. Fungi may appear Gram positive or Gram negative, but this reaction is due to the thickness of the cell wall, which either allows diffusion of the initial stain to enter in the cell or blocks its entry altogether. The thickness of the cell wall in fungi and other organisms has been suggested as a parameter that renders an organism as a Gram positive or Gram negative reaction.

The fungal cell wall has a different chemical composition than bacterial cells. Fungi have a well-defined nucleus. Since the precise pH of the fungal cell is unknown, one could hypothesize that the fungal cell environment may not be acidic enough to allow basic dyes to bind to the cellular components. Fungi may not be truly categorized as Gram positive or negative, since the observed reaction may not be due to the binding of the crystal violet-iodine complex to the cellular components inside the fungal cell but is held back due to the lack of escape route. The observed reactions are likely to occur, since the thick fungal cell wall would either prohibit or permit diffusion of the Gram stain reagents across in both directions. Typically, yeast appears Gram positive, and filamentous fungi display Gram negative reaction. The primary cellular difference causing variable Gram reaction is the thick cell wall as well as the morphological structures of yeasts and filamentous fungi.

Yeasts are unicellular and have a much thicker cell wall than Gram positive bacteria. Yeast cells allow crystal violet to enter the cell, followed by iodine. During the decolorization step, the decolorizer is not left on the smear long enough to damage the thick yeast cell wall. As a result, the crystal violet-iodine complex is retained inside behind the thick cell wall of yeast. However, the purple color in yeast is not necessarily an indication that the crystal violet is bound to the internal cellular material. Interestingly, if smears are air-dried and unfixed, yeast cells appear Gram positive, but when excessive heat is used to fix smears, yeasts usually appear Gram negative. Obviously, excessive heat damages the cell wall enough to allow CV to escape. This suggests that the damage to the yeast cell wall is extensive, since the decolorizer causes more damage to the cell wall when the smears are excessively heat fixed.[7] When decolorizer is applied only for few seconds to the gently heat-fixed smear, it causes little or no damage to the thick cell wall of yeast cells. Therefore thick-walled yeast cells appear purple and are assumed to be Gram positive.

Filamentous fungi are multicellular. Fungal hyphae (septate and aseptate) frequently appear pink (or Gram negative) when stained by Gram's method. The resultant pink color of fungal hyphae does not indicate that true binding of safranin has taken place in the interior of the cell. The cell wall of multicellular hyphae has a much thicker cell wall than the yeast cell wall. Fungal hyphae are often suspended in colloidal tissue material (during infectious process in the tissue). Therefore, Gram-stained fungal hyphae often display pinkish dye concentrating around the outer boundary of the cell wall while the interior of the cell remains unstained.

As a result of the apparent non-specific staining, tracks of fungal hyphae are frequently missed in Gram stain but give a clue to follow the morphology of the hollowness of the hyphal tracks developed by the contrast produced by safranin surrounding fungal elements. It is often found useful upon review of the smear subsequent to the recovery of fungus from the corresponding culture that the fungal elements appear obvious. Therefore, a majority of the time, the primary stain CV does not enter the fungal hyphae but is retained outside. The use of decolorizer damages the cell wall to a degree that allows the counterstain safranin to diffuse inside some of the fungal hyphae, staining pink, and it may be assumed to be Gram negative.

It is technically incorrect to call fungi Gram positive or Gram negative, since fungi are not meant to be categorized by Gram reaction; although they do show purple and pink color at times. Fungi are classified according to the morphology of the structure and are not based on Gram reaction. However, the Gram stain is an extremely useful procedure to detect the presence of fungal elements, regardless of their Gram reaction (color).

Human (animal) cells are also unable to be differentiated by the Gram stain reaction. Human cell does not have a cell wall; therefore human cell is not able to hold onto the basic dye used in the Gram stain, and it is immediately removed from the cell upon using the decolorizer that was a disappointment for Hans Gram when he tried to stain tubular cast from human cell line. However, Hans Gram's disappointment led to the accidental discovery of the Gram stain procedure that still holds valid and would remain an exclusive entity to exist as an *icon* within microbiology. ***Gram's method would now be seen as even more useful for the first time,*** *not only for the detection of bacteria but also for the detection of other organisms, specifically the fungi that were not indicated for detection by using the simple procedure of Gram stain.*

2.7 Why are yeast Gram positive and true hyphae Gram negative?

Hans Gram perhaps may have never thought how his method of Gram stain would be interpreted. Should we interpret the color purple as Gram positive and pink Gram negative or the binding of CV dye to the negatively charged cellular material inside the cell behind the cell wall? The author's experience suggests that both explanations may be true. A majority of the yeast cells appear Gram positive since they frequently stain purple, and a majority of true hyphae staining pink interpreted as Gram negative explain that the thickness of the cell wall in yeasts retaining CV-iodine complex behind the cell wall does not get damaged enough to produce large pores in the cell wall during the brief application of decolorizer. As a result, yeast cells appear purple, but not necessarily because they are truly Gram positive. There is no binding of the cation dye to the interior component of fungal cells, but they still show purple color due to the fact that the thick cell wall simply traps the large complex of CV-iodine, disallowing it from escaping after the decolorization step because the shorter period applied to the thick cell is not sufficient to make larger pores in the thick cell wall for CV-iodine to pass through. For this reason, the yeasts stain purple; however, they may not be truly Gram positive, since no chemistry plays a role inside the yeast cell.

The fungal hyphae on the other hand, having a much thicker cell wall, may or may not allow the diffusion of CV inside the fungal hyphae. If CV enters the fungal cell slowly, the iodine would follow the same route to form a complex with CV. After the decolorization step, the fungal hyphae would behave the same way as yeast, displaying purple color since the trapped CV-iodine is held behind the cell wall. However, the greater thickness of the fungal cell, the true hyphae, usually do not allow entry of CV (the most colorful and important primary stain). Iodine applied in the next step, which has no value without the CV in the Gram procedure, also remains outside the fungal hyphae. During the decolorization step, the decolorizer may be able to cause damage to the cell wall to a degree that would be sufficient to allow safranin to get in as opposed to the initial state when the thick cell wall of true hyphae disallowed CV to enter the cell. As a result the true hyphae may be stained pink and appear Gram negative or remain unstained when no reagent is allowed to enter inside the fungal hyphae before or after the use of decolorizer.

The author re-evaluated the simple test to demonstrate the role of the thickness of the cell wall, explaining why yeasts usually stain Gram positive and hyphae Gram negative. Two sets of smears were prepared using *Staphylococcus aureus*, fungal hyphae, and yeast. One set of the smears was stained by regular Gram method by flooding CV followed by iodine, decolorizer, and safranin. For the second set of smears, **CV and iodine were mixed** together before being applied to the smears followed by decolorizer and safranin. Results were predictable: The first set of smears stained by the regular Gram method displayed *Staphylococcus aureus* and yeast as Gram positive and hyphae stained Gram negative. The second set of smears displayed *Staphylococcus aureus*, fungal hyphae, and yeast; ***all three cell types*** stained pink as Gram negative (*Image set 2.7*).

Image Set 2.7

Stained by Regular Gram Stain Procedure	**Stained initially by CV-iodine mixture**

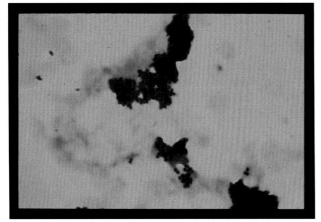

Gram stain x1000 (*Staph aureus*)

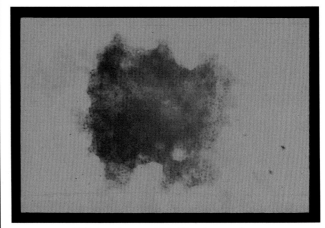

CV & iodine x1000 (*Staph aureus*)

Gram stain x1000 (yeast)

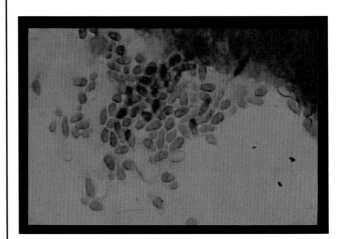

CV & iodine x1000 (yeast)

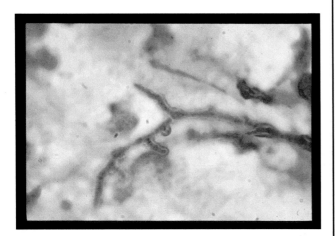

Gram stain x1000 (septate hyphae)

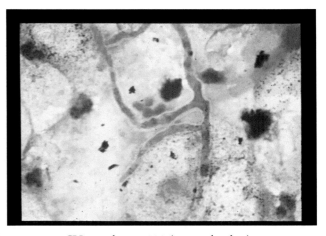

CV & iodine x1000 (septate hyphae)

The results from the above experiment prompted two questions: 1) Is yeast showing Gram positive reaction due to the thick cell wall? 2) does the same rule apply to the hyphae (filamentous form)? As mentioned before, the interior of the yeast cell is not acidic in nature; therefore the greater degree of CV would not bind to the cellular components in the interior of the yeast cell after CV gains entry. The true fungal hyphae remain pink; i.e., Gram negative in both sets of smears stained by two different methods. Looking at the results it may be reasonable to say that the cell wall of the fungal hyphae has far greater thickness as compared to the yeast cell wall. The fungal hyphae simply did not allow the diffusion of the CV within the time frame set in the Gram stain procedure. However, safranin may have difficulty in entering the cell of true hyphae. As a result, it stays outside of the cell wall boundary. Safranin may be able to enter inside the fungal hyphae and stain pink if sufficient damage to the fungal hyphae has been done by the decolorizer.

Application of the decolorizer causes some degree of damage to the cell wall that easily allows safranin entry to some hyphae, or it is blocked from entering the cell if the decolorizer has not been able to cause sufficient damage to the fungal cell wall. In this case, the fungal hyphae would not stain at all but would show unstained tracks of fungal hyphae, making the detection difficult when examined under the microscope. At this point, safranin plays an excellent role in making the detection easier when its deposits remain attached to the outer walls of fungal hyphae, making those canals like unstained tracks easier to visualize when observed microscopically.

This hypothesis has been proven true by the fact that *Staphylococcus aureus*, fungal hyphae, and the yeast cells staining pink (Gram negative) in the second set when stained by the mixture of the suspension of CV and iodine combined (instead of CV alone) failed to enter the cell wall due to the fact that larger molecules of CV and iodine complex could not pass through the thick cell wall (in both directions).

In conclusion, CV-iodine complexes, when formed within the cell, do not get easily removed from the yeast and fungal hyphae. The same is true when CV-iodine complexes are formed outside the cell wall of yeast and fungal hyphae; the large molecules formed are prevented from entering the thick cell wall of yeast cells and fungal hyphae. Therefore yeasts and fungal hyphae would stain variably depending on the degree of damage occurring to the cell wall before or after the use of decolorizer. If the CV-iodine complex is formed in the interior of the cell, the purple color reaction would have been hindered inside the cell environment and unable to escape, giving Gram positive reaction to the fungal hyphae. On their own, basic dyes (CV and safranin) are able to diffuse inside the cell. The fungal hyphae having too thick a cell wall would not allow CV to diffuse within the shorter time allowed to stain bacteria only.

The author's hypothesis suggests that the statement *"fungi are Gram positive"* mentioned in the microbiology and mycology textbooks is absolutely incorrect for two reasons: 1) Fungal cellular material is not acidic; therefore binding of the basic dyes does not take place inside the fungal cell behind the cell wall; and 2) the color purple is not an appropriate basis to call an organism Gram positive. If so then all Gram positive organisms that naturally stain Gram negative should be removed from the Gram positive category.[9, 10]

Hans Gram's intention was to detect bacteria using basic dyes that helped divide bacteria into two distinct groups based on the chemical reaction within the cell; some turned purple and others pink. The Gram reaction was not exclusively related to the physical properties of the cell wall of bacteria but the chemistry that played a role behind the cell wall of bacteria. Rigid and much thicker cell walls of fungi play a direct role in turning yeasts and mycelium purple or pink when stained by Gram's method. However, the color production is not related to the internal chemistry inside the fungal cell wall as it occurs in bacterial cells. It is for this reason that fungi are neither Gram positive nor Gram negative, although they do show purple or pink color. They are identified without making any distinction based on the color they produce, since fungi, whether Gram positive or Gram negative or remaining unstained altogether, would not require altering the antifungal therapy of the patient. On the other hand, bacterial classification based on the Gram reaction would require immediate action to change the antibacterial therapy if empirical treatment by antibiotics is found inappropriate after knowing the Gram reaction of the infectious agent.

2.8 UTILITY OF THE GRAM STAIN FOR FUNGAL DETECTION

The Gram stain is a simple and rapid procedure used in all microbiology laboratories to categorize bacteria into two distinct groups such as Gram positive and Gram negative. This initial differentiation from direct smears aids in the selection of empiric therapy for patients suffering from bacterial infections. The Gram stain technique is widely considered a bacteriological tool; however, it also has the potential to be utilized in mycology. Unfortunately, there has been no strong movement to use the Gram stain for fungal detection. Instead of describing the utility of the Gram stain for fungal detection, the medical mycology or microbiology textbooks actually discourage its use. This results in the loss of a great opportunity to influence patient care, particularly when a fungal infection is unsuspected.

Some microbiology laboratories maintain a separate mycology section in which medical technologists with significant mycology experience are employed. However, not all microbiology laboratories have a separate mycology section, and many times fungal samples are processed by technologists without significant mycology experience. Accordingly, it is essential to utilize techniques with which all technologists are comfortable for the early detection of fungal elements.

One important alteration in laboratory procedures to improve fungal detection in direct smears involves the rejection criteria for sputum specimens. The majority of fungal infections are acquired through inhalation. Therefore, the lung is the primary focus for fungal infections. It is common practice in the routine microbiology laboratories to screen sputum specimens to determine the quality of the specimen before it is processed for bacterial analysis. A sputum specimen is often considered unsuitable if it has greater than 25 epithelial cells under a low-power (x100) objective lens.[3] If a fungal culture has not been requested, this sample will be rejected without further analysis. Therefore, smear readers must be vigilant and consider the potential presence of fungal elements when reading direct smears, even if a fungal culture has not been requested. A specimen rejected for culture and susceptibility must not be discarded if fungi are observed in direct smear. In the event that fungi have been detected in the direct sputum specimen that has been rejected based on Q-score, it should be processed for fungal culture even if the specimen has not been specifically requested by the physician for fungal investigation.

2.9 COMMON GRAM STAIN ERRORS DO NOT AFFECT FUNGAL RESULTS

The Gram stain reaction is dependent on the strength of the crystal violet and iodine binding.[1, 4, 39] The binding of the CV-iodine complex to the bacterial cell wall is essential for Gram positive and Gram negative differentiation, upon addition of the decolorizer.[1, 4, 39] Many components of the Gram stain reaction can lead to over-decolorization and the identification of a Gram positive organism as Gram negative. These features include a low concentration of crystal violet, fading iodine, excessive washing after the addition of crystal violet, the use of too much decolorizer, and leaving the counterstain on the smear too long.[1, 4, 39] These technical errors may result in a false Gram stain report for bacterial classification but do not affect the identification of fungal elements. As fungi are not classified into groups by the Gram stain reaction, it is solely the detection of fungal elements in the smear, not the resultant color, that is the key to report to the clinician to aid patient care.

Over-decolorized and under-decolorized Gram smears would provide erroneous and misleading interpretation for bacteria if such Gram reaction is not carefully observed by the smear reader in the microbiology laboratory. An under-decolorized Gram smear reaction would cause an interpretive problem for fungal detection. Under-decolorized smears may hide fungi beneath the precipitated Gram stain reagents, making the detection of fungal elements very difficult. However, over-decolorized Gram-stained smears are less likely to cause an interpretive problem, since Gram reaction is more important for the detection of fungal elements regardless of what color outfit they wear.

2.10 REPORTING OF FUNGAL ELEMENTS SEEN IN DIRECT SMEARS

The most important component of detecting fungal elements in direct smears is the complete and accurate description of the observed structures. First, identify whether oval/spherical or filamentous form are observed.

Flowchart 2 is designed to aid the observer in differentiating round/oval cells. If yeasts are seen, the first differentiating feature is whether or not budding is observed. The second step in differentiation is to determine if the cells are small or large, round or oval, thick walled or double walled, with either a narrow or broad-base attachment. The observation of budding forms is essential to reporting the observed form as a yeast. The non-budding cells resembling yeast should have their structures defined without elaborating further identification, as specific identifications cannot be made from the direct smear. Care must be taken to observe round and spherical cells that are not fungi but morphologically resemble yeast or yeast-like fungi. Any microscopic structure resembling yeast must be carefully observed for budding. Any round or oval cell resembling yeast should not be called a yeast unless budding is demonstrated at some point under the microscope. There are structures that resemble yeast or other fungi; however, they are non-budding. Some are fungi, and some others belong to non-fungal entities.

If fungal filaments are seen in the direct smear, determine if the filaments are <1μm or >1 μm in width in order to divide organisms into one of the two categories: bacillary form and fungal form. Observe filaments for branching and non-branching to separate actinomycetes from mycobacteria. If fungal hyphae are seen, pseudohyphae and true hyphae must be differentiated. This can be done by examining each segment of the filament (pseudohyphae) and observing constrictions at the ends. If true hyphae are seen with irregular width and sparse septation and appear as a ribbon-type—too wide in some areas and narrow at the other end—these are usually reported as aseptate hyphae. If true hyphae are observed with frequent septation within the short length of a filament, parallel filament sides, and without constrictions at the joining ends, these are identified as septate hyphae. In some cases, the organism may display clear dichotomous branching.

Additional information may be available from the smear such as the pigmented or hyaline nature of the septate hyphae in the Gram smear (or KOH preparation). Unfortunately, dematiaceous fungi do not always demonstrate pigment using routine microbiology-staining procedures, such as the Gram stain. Therefore, a special stain such as Fontana Masson to detect the melanin pigment present in the fungal cell wall of dematiaceous fungi is most useful for this purpose (refer to Figure 1). Some fungi in the dematiaceous group contain a lesser amount of melanin pigment not usually observed by wet prep or Gram smear. However, dark fungi often provide some clues to suspect them, such as cells in chains forming wider attachment at the segments or regular swellings (caterpillar type) along the cells in chains (toruloid structure) looking like pseudohyphae. However, such structures are usually seen as non-budding, and marked septum is observed in between the segments instead of pinched-off constrictions that are commonly observed in pseudohyphae.

Other such examples are "copper pennies," "muriform," or "sclerotic bodies," indicating dark fungus involved in the site-specific fungal infection known as Chromoblastomycosis.

2.11 APPROPRIATE WORDING FOR FUNGI DETECTED IN DIRECT SMEARS

When reporting fungi detected in a direct smear, the clinician should be given as much detail as possible. "Fungal elements seen" is a vague term frequently utilized as it is sometimes difficult to further interpret the observed fungi. The term "fungal elements seen" should be restricted to cases in which a more accurate description of the fungi is not possible.

2.12 NEGATIVE REPORT

- No fungal elements seen

2.13 POSITIVE REPORT

Use one or more of the terms below as applicable:

- Budding yeast seen
- Budding yeast with pseudohyphae seen
- Fungal elements seen
- Round or oval yeast cells with unipolar bud, consistent with *Malassezia* species, a normal skin flora
- Round yeast cells and short filamentous pseudohyphae ("meatballs and spaghetti"), consistent with Pityriasis
- *Pneumocystis jiroveci* seen
- *Pneumocystis jiroveci* seen; no other fungal elements seen
- Septate hyphae seen
- Septate hyphae seen consistent with dematiaceous group (dark pigmented)
- Aseptate hyphae seen, suggestive of *Zygomycetes* (mucormycosis)
- Large, round, double-walled budding yeast with broad base, suggestive of *Blastomyces dermatitidis*

Other structures seen must be defined based on the shape and size with or without suggesting the name of the organism.

2.14 UNSPECIFIED NICHE OF GRAM-STAINED SMEAR IN MYCOLOGY

Gram smear is not an excellent procedure for the detection of fungal elements in the clinical specimens. However, recovering fungi in the Gram-stained smear prepared from clinical specimen, especially when the physician least expected it in his/her patient, puts the simple Gram-stained smear well beyond the excellency mark. Therefore, the question about the importance of the simple test is not as important as the intent and the effort of a medical technologist to find something without knowing what before hitting the jackpot; i.e., recognition of fungi in the Gram smear.

CHAPTER 3

3.1 CLINICAL CASES FOR FUNGAL DETECTION IN DIRECT SMEARS

Discussed below are cases that highlight some of the difficulties encountered by technologists during microscopic examination of Gram smears and suggestions to ameliorate these problems. The major problems that are frequently experienced with the Gram stain in routine microbiology are under-decolorized and over-decolorized smears. Fungi do not usually get affected by under- or over-decolorization reaction; however, they come across problems in detection due to unstained, buried-under-debris, and distorted fungi.

3.2 DIMORPHIC FUNGI

Dimorphism in fungi in simple terms is the physical structural changes taking place in fungi growing in different environment, such as mycelial growth occurring at room temperature in a natural environment (soil, wood) and yeast, yeast-like, and spherule phase in a different environment at 37^0C in the incubator or human body temperature. Since fungi feed on decaying organic matter, they do not need to depend on a human or animal host for nutritional requirements; therefore, they may not be able to invade mammalian tissue. The fact that fungi are restricted in causing infections in humans is heavily based on three main factors: temperature, host immunity, and nutritional requirements. In order for fungi to feed on healthy human tissue, they need to adapt to a human tissue environment. It is like fungi are asking humans, "If you want us to come to your home (host tissue), make some changes at your end and we would also bring some changes with us." Fungi bring structural changes within themselves to be able to survive in the host environment, and humans change a little bit by becoming immunocompromised either by a cause (immunodeficiency, malignancy, chemotherapy, etc.) or on purpose (organ transplant). In other words, all fungi would change physically in order to be able to get established in human tissue.[20, 39]

The dimorphic fungi grow in two phases: They grow in mycelial phase at room temperature or at 25^0C to 30^0C; and in yeast or spherule phase in the host tissue or at laboratory temperature 37^0C. The majority of dimorphic fungi cause systemic mycosis in immunocompromised patients spreading to other organs via metastasis. Some dimorphic fungi cause mild infection that resolves spontaneously and may go dormant and cause systemic infection later on. The most commonly encountered dimorphic fungi are *Histoplasma, Blastomyces, Coccidioides, Sporothrix, Paracoccidioides, Penicillium marneffei* and *Emmonsia parva* (adiaconidia).[39] In order to acquire infections by dimorphic fungi, conidia or mycelium are usually inhaled, and the process of infection begins within the host tissue whose immune system has been depleted either temporarily or permanently due to disease, organ transplant, or other reasons. The yeast phase of dimorphic fungi does not initiate infections. Therefore, the yeast phase must convert to the mycelial phase before being able to infect the next vulnerable victim.

For practical purposes, it is important for clinical microbiology to accurately define structures and be able to interpret them as fungi in the Gram-stained smear, since many times clinical specimens received by microbiology are asked for C&S only and not for fungus. Following are some cases that reflect the scenario about missing fungi in the specimen that remained unreported. The following are examples of dimorphics that were suspected or detected in the Gram-stained specimens collected from clinical specimens.

3.3 Blastomyces[8, 9, 20, 23, 29, 33, 42]

Blastomyces dermatitidis, a dimorphic fungus, causes acute, chronic, granulomatous infections of the respiratory system and may spread to cause systemic mycosis. It is most commonly found in North America, Mexico, Central America, and Africa. The largest number of cases originated from the Mississippi, Ohio, and Missouri river valley regions. Blastomycosis is more common in men associated with outdoor activities. *Blastomyces* is a dimorphic fungus. It appears in the yeast phase in human tissue (body temperature) or at 37^0C in laboratory environments and grows in mycelial phase at room temperature ($< 30^0$C). In clinical specimens, *Blastomyces* cells appear in the yeast phase as round, double-walled with broad base (ranging 4–5 μm) cells ranging from 3μm to 30μm (average size being 8–15μm). There are two cell-sized variations: as small and large, both of which have the same microscopic morphology features.

Technically, the recognition of *Blastomyces* is straightforward. However, in real-case scenarios, the textbook description is not always available. During such times when decision making is difficult for one of several reasons, the laboratory staff in clinical microbiology must develop some sort of aptitude for the recognition of unusual cells and know about their important features so that the organisms when they appear distorted or undifferentiable may be identified by comparison with similar-looking organisms falling in the same size and shape. Using the flow chart provided, develop a technique to rule out one organism at a time to reach the final stage of decision making. In case the identity of the organism is unclear, try to define structures seen under the microscope about the fungus in question without elaborating any further and forward the specimen or smear to the mycology experts and follow the referred report when available to learn for the next time around. The smear readers have to learn to question themselves in order to generate ideas and expand them during the self-questioning and answering mode.

Blastomyces dermatitidis is a health-hazard fungus. Care must be taken to work with this organism using proper safety procedures such as working in the bio-safety cabinet and level 2.5 to 3 lab safety containment facilities. *Blastomyces* grows slowly as white (some *Blastomyces* grow faster) and cultures are held for a full four weeks at 30^0C; if direct smear shows the evidence of *Blastomyces* or it is requested specifically by the clinician, the negative culture media must be incubated for six weeks.

3.4 Case 1

A male patient was treated for lung cancer. A BAL specimen was sent to the microbiology laboratory for culture and susceptibility and fungal analysis. No fungal elements were seen in the initial Gram stain or fungal smears. About 10 days later, a dull type of fungus grew on mycology media. A lactophenol cotton blue (LPAB) preparation identified the fungus as *Blastomyces dermatitidis*.

Upon isolating *Blastomyces*, the Gram smear and the fungal smears were reviewed. Rare, round, medium-sized single cells were seen in the CW. However, the observed structures were nonspecific, not having enough information to call the objects *Blastomyces*. The corresponding Gram smear revealed double-walled, broad-base budding cells (*Image set 3.4*). Based on the characteristic morphology, the clinician was notified of the *Blastomyces* present in the BAL specimen. *Blastomycosis* was an unexpected diagnosis, and the direct smear results had an immediate impact on patient care.

As described in Flowchart 2, *Blastomyces* should only be suspected based on the observation of a single cell by CW. The confirmation of *Blastomyces* by microscopy requires the detection of the broad-based budding structure, which was not present in this case. *Blastomyces* is a large, round, double and thick-walled budding yeast with broad base (8–10 μm). When a few cells are observed that have some features of *Blastomyces*, it is necessary to consider all cells with similar microscopic morphology for defining characteristics. *Blastomyces*, *Cryptococcus*, *Pneumocystis*, *Histoplasma duboisii*, myospherule, adiaconidia, and *Loboa loboi* have round or near spherical morphology; therefore, they must be differentiated based on their size, shape, and specific details.

The broad-base structure in *Blastomyces* begins to appear during the developmental stage. The mother cell acquiring maturity, a bud is formed to give rise to a daughter cell that remains attached to the mother cell until it matures and is separated. The broad-base bud is developed between the two cells, allowing the transfer of cytoplasmic material from the mother cell to the newly formed daughter cell. *Blastomyces* has a double wall; however, smear stained by CW would only show thickness in the cell wall without revealing separation between the cell wall and the cytoplasmic membrane under the UV light. The Gram-stained smear at this point would serve a very useful role in demonstrating not only the thick wall of the *Blastomyces* but also indicating the separation of cell wall and cell membrane known as "broad-base budding," a characteristic that is unique for *Blastomyces dermatitidis*.

Blastomyces, when present in its perfect state resembling the textbook description, is easily identified. However, it is often difficult to recognize *Blastomyces* when it is not in its picture-perfect recognizable state. Missing *Blastomyces* due to distorted structures would have enormous adverse impact on the patient as well as misleading the clinician for correct diagnosis and treating the patient appropriately. In another case, a BAL specimen from a sick patient suffering from pneumonia was received by microbiology for routine C&S viral studies, TB culture, and fungal investigation. The patient was diagnosed and was on treatment for cancer-related illnesses. All other microbiology tests were negative, except that weird structures of *Blastomyces* (**Image in set 3.4**) were seen under the UV microscope in CW prep. It was a total surprise for the clinician to be notified of the presence of *Blastomyces* in the patient's BAL specimen, and he came down to the lab to examine the fungal smear himself. The structures seen under the microscope were so different that the clinician was unable to confirm the structures seen as *Blastomyces*. The clinician was unimpressed by what was seen under the microscope but was also careful not to disagree with the mycology expert reporting the structures as *Blastomyces*, and he commented about the author having "a weird imagination." About two weeks, later a white mold was isolated and identified as *Blastomyces dermatitidis*.

Image set 3.4

3.4 A BAL: Gram x1000 bizarre *Blastomyces dermatitidis*

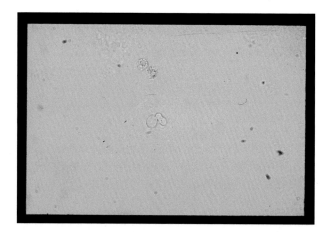

3.4-B BAL: WP x400 bizarre Blastomyces dermatitidis

3.5 CASE 2

A post-lung-transplant female patient was admitted to the hospital for routine follow-up procedures. Her BAL sputum specimen was collected and sent to the lab for microbiological analysis, including fungal investigation. The specimen was stained with CW in mycology and examined under the UV microscope. Many round structures of variable size were seen. At some point they appeared like immature and mature spherules of *Coccidioides* but were also showing broad-base budding mimicking *Blastomyces* (**Image set 3.5**). About 10 days later *Blastomyces dermatitidis* was recovered from the culture. No *Coccidioides* was isolated. The clinician did not suspect *Coccidioides*, but *Blastomyces* and the patient's travel history did not support the presence of *Coccidioides* in her BAL specimen.

Often, the morphology of the normal cells is altered due to the physical and mechanical manipulation of the specimen containing fragile organisms whose structures are vulnerable to pressure and as a result cause distortion to the delicate morphology of the organisms. The reason for suspecting structures resembling *Coccidioides* was that the cells burst open, displaying characteristic images of *Coccidioides* spherules. The cell wall of the single cells showing thick wall was another clue in suspecting *Coccidioides*, immature spherules, and not *Blastomyces*. However, it was necessary to examine the structures more closely in order to be able to interpret objects seen under the microscope correctly. Since there was overlapping morphology of two dimorphic fungi, the patient's travel history and other data were needed to reach a successful conclusion. This helps to determine the clinical significance by establishing direct communication between the microbiology laboratory and the clinician. After lengthy discussion with the clinician, the organism was reported as *Blastomyces*, ruling out *Coccidioides*. The corresponding culture grew a white mold, identified as *Blastomyces dermatitidis* and confirmed by morphological and DNA studies.

Image set 3.5

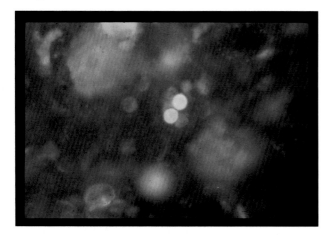

3.5-A BAL: CW x400 single *Blastomyces dermatitidis*

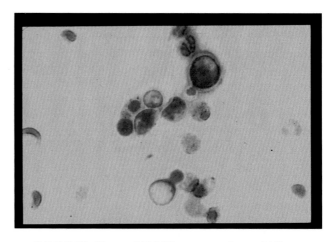

3.5-B BAL: Gram x1000 Blastomyces dermatitidis
(*double wall and broad base*)

3.6 CASE 3

A 75-year-old male patient (immunocompromised) was admitted to the hospital for respiratory complications. A BAL specimen was submitted to the microbiology department for C&S and fungal investigation. The Gram smear results were reported as "no pus cells or bacteria seen." Fungal smear stained by CW showed large round double-walled budding yeast with broad base consistent with *Blastomyces*. The Gram smear was reviewed and *Blastomyces* cells were found in the Gram-stained smear, but the smear was *grossly under-decolorized* (**Image set 3.6**). The under-decolorized Gram smear *was re-stained by using decolorizer* followed by safranin. Structures that were previously buried under the stain deposits were clearly visible, with an excellent textbook description of *Blastomyces*. The culture grew *Blastomyces* in corresponding culture media in mycology.

The question comes to mind, why did we decide to review the Gram smear, since the fungal smear (CW) and wet prep were both loaded with *Blastomyces*? The diagnosis of the fungal infection had already been made. However, *Blastomyces* being large in size and present in greater quantity was missed in the Gram-stained smear, and that made us wonder why this fungus remained undetected. It was just a proficiency issue to find out what went wrong and why. There is no way that large *Blastomyces* cells could have been missed by the smear reader. Upon review, we noticed two important irregularities in the process. The Gram smear was under-decolorized and absolutely unsuitable for examination. The technologist reading the under-decolorized smear did not raise the issue to the technical staff responsible for making Gram smears from the processed clinical specimens and did not reject the defective Gram-stained smear. The fact that the Gram smear reader missed *Blastomyces* was due to the poor quality of the smear that resulted in the *Blastomyces* being buried under the Gram reagents, specimen material, and debris. The Gram smear should have been rejected and a replacement requested. Or the original under-decolorized smear should have been re-decolorized followed by safranin application.

This example signifies the importance of the quality of smears prepared. Rushing to make simple Gram smears may result in skipping steps either in the rush or in knowing the procedure too well. As mentioned earlier, many times clinical specimens received in microbiology do not request fungal investigation. During these times, Gram-stained smears serve as excellent alternatives for finding fungi in clinical specimens. The poor quality of Gram-smear-suppressing *Blastomyces* is another reason for the technologists to be aware of the fact that suboptimal smears must be rejected and a repeat requested for replacing the poor-quality Gram-stained smears. Poor-quality Gram smears also produce suboptimal results for bacteria.

Image set 3.6

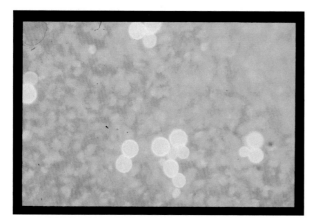

3.6-A BAL: CW x400 *Blastomyces* -
missed in under decolorized Gram

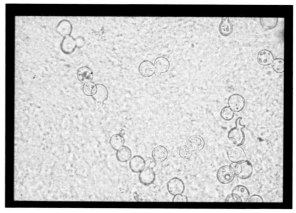

3.6-B BAL: WP x400 Blastomyces -
missed in under decolorized Gram smear

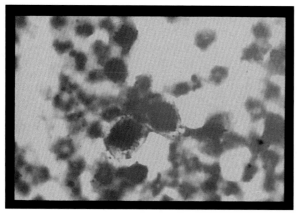
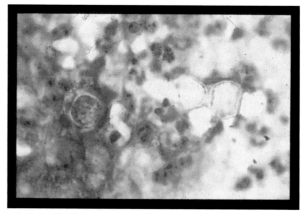

3.6-C BAL: Gram x1000 *Blastomyces -* missed in under decolorized Gram

3.6-D BAL: Gram re-stained x1000 *Blastomyces dermatitidis*

3.7 CASE 4

A 44-year-old male lung biopsy specimen was received by the microbiology department for C&S, viral, TB, and fungal analysis. Gram stain results reported as "moderate amount of pus cells and NBS." No growth was obtained from C&S culture. TB and viral study were negative. The fungal smear stained by Fungi-Fluor (CW) examined under the UV microscope showed very few medium-sized, single, round yeast-looking cells that appeared to be having double wall consistent with *Blastomyces*. The Gram smear was reviewed and occasional distorted cells were seen that needed further work to make these cells stand out clearly. The Gram smear was overstained by ZN, since ZN stain is excellent to stain *Blastomyces* (Image set 3.7).[5] A Gram smear overstained by ZN did not provide contrast, but the morphology of *Blastomyces* was little better than seen in Gram and fungal-stained smears. At this point the original Gram was overstained with KOH/CW and examined under the UV microscope as well as under the bright-field microscope. Excellent structures of *Blastomyces* were now seen under UV and bright field displaying large, round, double-walled budding yeast cells with broad base. Fungal culture grew white mold in about seven days, identified and confirmed as *Blastomyces dermatitidis*.

This case signifies the need to make extra efforts to detect fungal organism by making extra smears for staining or using overstaining technique on Gram smears (see the appendix for methods). Often, only a few cells are present in the clinical specimen to begin with, some of which may be hidden under specimen debris or staining reagents; others may get damaged by mechanical manipulation and appear distorted. Unusual structures pose problems for correct identification. Laboratory technologists must be prepared to take some extra responsibility by spending more time to review Gram smears, making more if necessary or overstaining to resolve the conflict, because any information reaching the clinician may prompt him or her to take immediate action to initiate the patient's therapy.

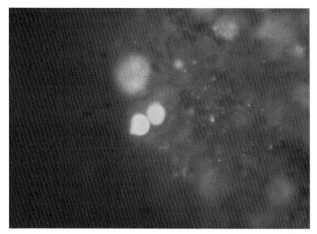

3.7-A Lung Biopsy: CW x400 *Blastomyces dermatitidis*

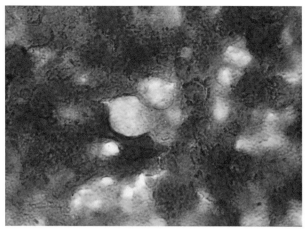

3.7-B Lung Biopsy: Gram x1000 *Blastomyces dermatitidis*

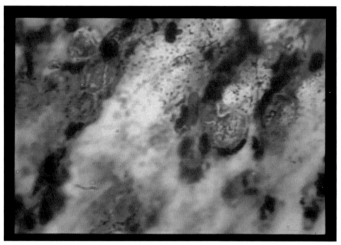

3.7-C Lung Biopsy: ZN x1000 *Blastomyces dermatitidis*

3.8 CASE 5

Brain abscess aspirate collected from a 55-year-old male patient revealed large round, double-walled budding yeast cells with broad base consistent with *Blastomyces* in fungal stain (CW) smear. The specimen was requested to have C&S and fungal culture. Moderate pus cells and NBS were reported in the Gram smear. No growth was obtained in C&S culture. Upon seeing *Blastomyces* in the fungal smear, the Gram smear was retrieved and reviewed; excellent *Blastomyces* were seen (*Image set 3.8*). Mycology culture grew white mold within four days, identified and confirmed as *Blastomyces dermatitidis*.

Image set 3.8

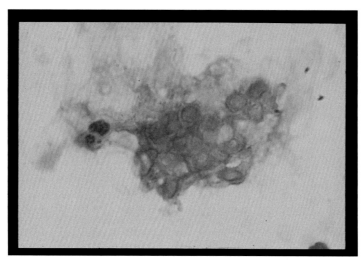

3.8 Brain Biopsy: Gram x1000 *Blastomyces dermatitidis*

3.9 CASE 6

A skin biopsy specimen collected from left forearm of an 87-year-old male patient was received in microbiology for C&S, fungal, and TB culture. The Gram stain contained scant PMNs; scant Gram positive and negative bacteria. Scant growth of commensal flora was reported from C&S culture. Culture was also negative for TB. Fungal stain was negative. About six days later a young colony of mold began to appear, which matured in a few days as white mold identified and confirmed as *Blastomyces dermatitidis*. Upon recovering mold from fungal culture, the fungal-stained smear was reviewed and no fungal elements were seen. A spare smear was stained with KOH/CW and viewed under the microscope (**Image set 3.9**). A single, large, round, double-wall budding yeast cell with broad base consistent with *Blastomyces* was seen.

Medical technologists reading Gram-stained smears often ignore or miss organisms other than bacteria since they focus so much on finding and interpreting bacterial identity, as that is the number-one priority in clinical microbiology. Microbiology technologists must be aware of the fact that any object seen under the microscope forming a specific morphology must be identified even if they are unable to identify it themselves. In these circumstances, the matter should be brought to the attention of the experts to resolve decision-making problems.

Image set 3.9

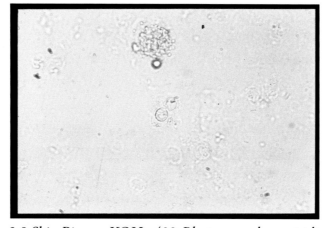

3.9 Skin Biopsy: KOH x400 *Blastomyces dermatitidis*

3.10 CASE 7

A 63-year-old male BAL specimen was received by the lab asking for C&S, mycobacterial, viral, PCP, and fungal analysis. Gram stain results reported as pus cells seen, NBS, and no growth from culture was reported. PCP, viral, and mycobacterial analysis were also negative. A fungal stain showed occasional large, round, and double-walled budding cells with broad base consistent with *Blastomyces*. A week later a white mold was isolated, identified, and confirmed as *Blastomyces dermatitidis*.

This particular case is important for two reasons. First, *Blastomyces* was called based on broad-based budding and double-walled structures. Second, a low number of *Blastomyces* not found in the Gram smear may have been missed for one of several reasons. Upon careful examination or review of the initial Gram smear, missing cells may have become obvious (*Image set 3.10*). Laboratory technologists must keep in mind the need to differentiate all cells seen under the microscope that have universal oval or round morphology and then gather as much information as possible to rule out one organism against the other based on species-specific clues and arrive successfully at the identity of the organism. For example, *Cryptococcus* cells may be as large as *Blastomyces* in a direct smear. Care must be taken to observe the thickness of the cell wall and the budding pattern of organisms. *Cryptococcus neoformans* producing large, round cells must be examined for narrow-base budding as opposed to thick and double-walled *Blastomyces* with broad-base budding. A most useful clue for *Cryptococcus neoformans* is the observation of dark cell walls under the microscope, indicating the presence of melanin.

Image set 3.10

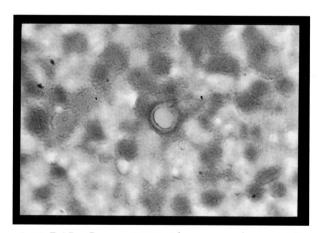

3.10 BAL: Gram x1000 *Blastomyces dermatitidis*

3.11 CASE 8

A 75-year-old male was admitted to the hospital with respiratory stress. A BAL specimen was collected and sent for microbial analysis. Gram stain reported as "no pus cells and NBS." However, white mold started to grow within three days in C&S. Viral study, PCP, and TB analyses were negative. The fungal smear showed excellent large, round, double-walled budding yeast with broad base consisted with *Blastomyces*. The Gram smear was reviewed, and *Blastomyces*-like cells were observed without a problem (Image set 3.11). White mold similar to C&S isolate was isolated within few days from fungal culture, identified as *Blastomyces dermatitidis*.

This example suggests that sometimes most easily observable *Blastomyces* are missed in the Gram smear due to the fact that the smear reader was focused on finding bacteria only and not thinking about *Blastomyces* cells that were present in every other microscopic field.

Image set 3.11

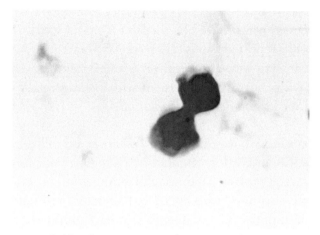

3.11 BAL: Gram x1000 *Blastomyces dermatitidis*
(note that this image does not demonstrate broad-base bud since it is about to separate from the mother cell).

3.12 CASE 9

A 44-year-old male BAL specimen was received for microbiology analysis. The Gram smear results reported as "pus cells seen, NBS," and culture results reported as commensal flora. All other tests, such as TB, viral, and PCP studies were negative. In mycology, fungal smear showed only three large, round, double-walled, and single (*Image set 3.12*), and few oval cells were seen. All cells seen were non-budding but appeared in yeast morphology. It was difficult to name these structures since no budding forms were seen. However, the yeast cells seen appeared to be *Blastomyces*. Fungal stain results were reported as yeast without elaborating any further. Six days later a white mold was recovered from fungal culture and identified and confirmed as *Blastomyces dermatitidis*. The Gram smear was reviewed and found few large, round, double-walled budding yeast with broad base. Recovery of *Blastomyces dermatitidis* was a total surprise for the clinician since the patient was suspected of having cancer.

Image set 3.12

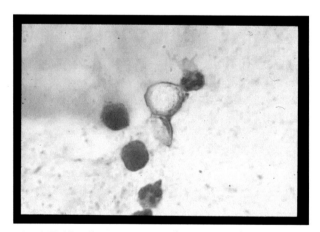

3.12 BAL: Gram x1000 *Blastomyces dermatitidis*

3.13 CASE 10

A sputum specimen was received by microbiology with a request STAT for *Nocardia*. Modified kinyoun stain was performed by the bacteriology staff and reported as negative for *Nocardia*. Since the patient was seriously ill and had symptoms suggesting nocardiosis; the Infectious Disease Services (IDS) team requested review of the modified kinyoun smear by the mycology expert. The Gram-stained smear and modified kinyoun smear were reviewed and found negative for *Nocardia*. However, there was something else seen in the smears that stunned the IDS team. Excellent structures of *Blastomyces* were seen in the Gram smear (Image set 3.13) and in the modified kinyoun smear.

It must be remembered that ZN or modified kinyoun stain is excellent for *Blastomyces*.[13] ZN stains provide better contrast to pick up *Blastomyces* easily. Double-wall and brood-base bud are seen with optimum clarity. This case emphasizes the importance of direct communication between the clinician and the laboratory staff dedicated to resolving the unusual results. The solution to this puzzle was very easy; however, it reiterates the philosophy of being aware of the structures that form a morphology and not overlooking or ignoring them because of lack of experience in mycology. Any object seen under the microscope presenting definitive morphology must be checked out or brought to the attention of an expert or the supervisor who would be in a position to make an appropriate decision to identify unknown structures seen under the microscope.

The author has added multiple images of *Blastomyces* from different specimens collected from patients with different clinical abnormalities for the reason that the classic textbook picture of *Blastomyces* is not always seen in reality. As shown in the image sets, *Blastomyces* appear in variable sizes and shapes; some of the organisms were not clear cut and have to be studied closely to interpret them correctly. One good thing about the discovery of *Blastomyces* in the Gram-stained smears is that it allows the medical microbiology staff to seriously consider the simple Gram-stain procedure a useful tool for finding fungi, especially when clinical request comes with the specimen for C&S only.

Image set 3.13

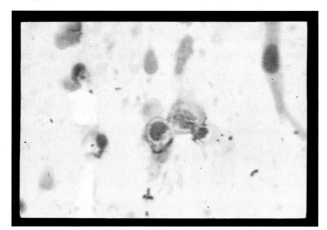

3.13-A Sputum: Gram x1000 *Blastomyces dermatitidis*

3.13-B Sputum (Culture): *Blastomyces dermatitidis* on IMA

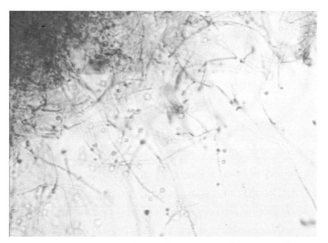

3.13-C LPAB x 400 *Blastomyces dermatitidis*

CHAPTER 4

4.1 HISTOPLASMA [8, 9, 20, 23, 29, 33, 42]

Histoplasmosis is usually contracted by inhalation. Therefore, mediastinal is the primary site for infection. *Histoplasma* conidia or hyphal fragments are inhaled and settle in the lungs. Severity of the infection is directly related to the inoculum size. Infections occur most commonly in the midwestern and southern parts of the United States. *Histoplasma* has also been isolated from other countries. In direct smears made from the clinical specimens, the yeast phase of *Histoplasma* appears as small oval budding yeast 2-4 μm with narrow-base budding, usually found intracellular. The size and the shape of *Histoplasma* overlap other yeasts to make conclusive evidence much harder when seen in respiratory specimens that frequently contain yeasts other than *Histoplasma*. Often *Histoplasma* does not display intracellular morphology in the clinical specimens; therefore, direct smear results must be carefully reported with descriptions and details of morphological features of yeasts observed under the microscope. The common etiological agent for histoplasmosis is *Histoplasma capsulatum var. capsulatum*.

Histoplasma slowly grows a white mold turning tan to light brown with age. Care must be taken to work with this level-three organism. Biosafety level-three lab containment facilities are required to work with *Histoplasma*. Extreme safety precautions must be carried out to work with this potentially hazardous fungus.

4.2 CASE 11

A middle-aged South American patient was admitted to the hospital for respiratory illness. Complete blood work, X-rays, and lung biopsy were ordered for laboratory testing to determine the cause of his respiratory distress. The lung biopsy specimen was processed for C&S and fungal investigation and NBS, and no fungal elements were reported from direct smears. The hematologist examining the blood film noticed cells resembling parasites and reported, "*Leishmania Donovani* (LD) bodies seen," to the clinician. The patient was put on anti-parasitic drugs to eliminate LD. However, the patient did not respond and continued to have symptoms. On day 11, a white mold appeared on culture, identified as *Histoplasma capsulatum* by microscopic morphology and confirmed by the reference laboratory. The patient was promptly treated with an anti-fungal agent and began to improve. He was discharged from the hospital within a few days.

Lots of homework took place after this episode of misidentification and inappropriate therapy to the patient. First, we asked ourselves where we erred in diagnosing structures as LD bodies and not *Histoplasma*. The suspicion of LD bodies in a hematology blood film was made by the hematologist based on the assumptions without consulting microbiology or histopathology. A Gram stain does not stain *Histoplasma* cells in the yeast phase as readily as it stains other yeasts; such poor stains require a trained eye and expertise to recognize them. The fungal smear stained by CW was also negative; however, upon re-staining the smear using small solid pieces sitting on the borders of the specimen area on the microscopic slide, small oval yeast with narrow-based budding were seen. The exact identity of the yeast seen under the microscope is not usually known, since several other yeasts overlap the size and the shape of *Histoplasma (Image set 4.2)*. Upon comparing LD bodies and *Histoplasma* stained by Giemsa, one may be confused, with the structures so closely resembling each other. Hematology should have sent the specimen to histopathology, where GMS is usually done to resolve fungal conflict. It must be remembered that Wright's and Giemsa's staining procedures are used in hematology and stain *Histoplasma* as well as leishmania (LD bodies). However, GMS would stain *Histoplasma* but not *Leishmania*.

Image set 4.2

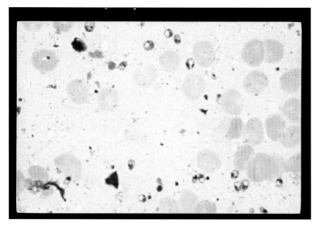

4.2-A Peripheral Blood: Wright's stain x1000
LD bodies - resembling *Histoplasma capsulatum*

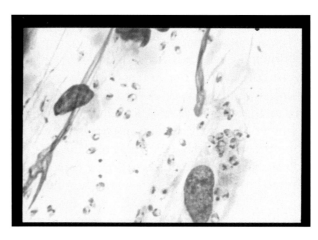

4.2-B Skin Biopsy: Giemsa x1000 LD bodies
resembling *Histoplasma capsulatum*

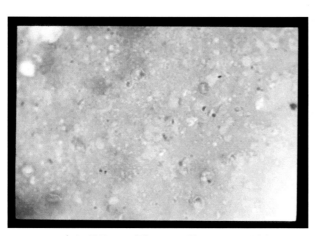

4.2-C Lung Tissue: Giemsa x1000
Histoplasma capsulatum

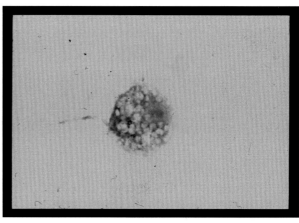

4.2-D BAL: Gram x1000 *Histoplasma capsulatum*

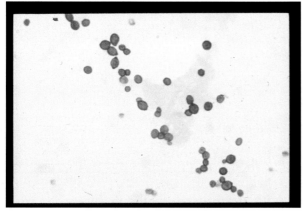

4.2-E Lung Biopsy: GMS x1000
Histoplasma capsulatum

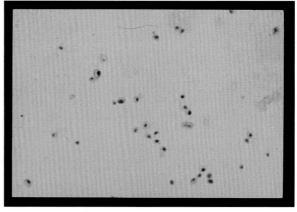

4.2-F Lymph Node: PAS x1000
Histoplasma capsulatum

4.3 CASE 12

A BAL specimen was received from a 32-year-old male patient suffering from AIDS. Clusters of cells in yeast morphology were observed in every other microscopic field in the fungal (CW) smear (A: *Image set 4.3*). Further differentiation of the cells could not be made from the CW. The initial Gram stain was reported as "no bacteria seen." However, the Gram stain was reviewed and yeast cells found that had previously been missed. The yeast cells were not well stained but were displaying pinkish and unstained hollow morphology located intracellularly (B: *Image set 4.3*). The dubious morphology of *Histoplasma* may have tricked the smear reader, and as a result *Histoplasma* was not picked up in the Gram-stained smear. The smear was difficult to differentiate from the surrounding cellular material. Based on the patient's clinical condition and the morphology of the intracellular yeast cells, a presumptive report of *Histoplasma* was issued. This identification was confirmed 10 days later when a white mold was isolated and identified (C: IMA and D: LPAB: *Image set 4.3*). Many times, a negative Gram smear result gives us an indirect clue for those fungi that do not stain normally by Gram stain procedure as other fungi do. Therefore an extra effort was needed to find out why clusters of yeast cells were seen in CW in every other field but the Gram remained negative. Two things turned out to be important in this case. Patient data from the clinician indicated the depleted immune status of the patient. The next thing that was needed was to know whether the organisms seen in the CW smear were intracellular or not. Since CW does not stain macrophages or PMNs that are easily seen in the Gram smears, the Gram needed to be checked once again to see if intracellular organisms were present in the smear or not. Upon reviewing the Gram smear, there was no doubt about the presence of *Histoplasma* in the patient's BAL specimen.

Image set 4.3

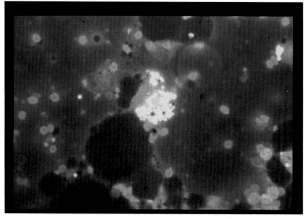

4.3-A BAL: CW x400 clusters of
Histoplasma capsulatum

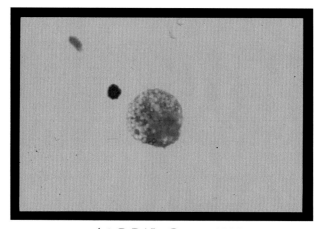

4.3-B BAL: Gram x1000
Histoplasma capsulatum

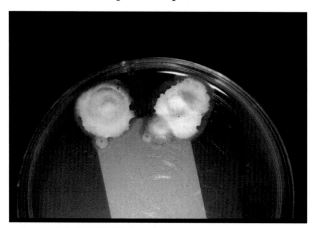

4.3-C BAL (Culture): IMA: *Histoplasma*
(white mould turning tan)

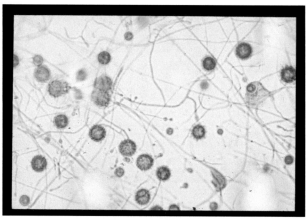

4.3-D BAL: LPAB x400 *Histoplasma capsulatum*
showing microconidia and tuberculate
macroconidia under the microscope

4.4 Case 13

A thirty-four-year-old male was admitted to the hospital, suspected of having pneumonia. A BAL specimen was sent to microbiology for C&S, TB, PCP, viral studies, and fungal culture. Gram smear results reported as "no pus cells seen, commensal flora present." All microbiology tests, other than fungal analysis requested, were negative. The fungal smear displayed small oval budding yeast seen, appearing to be located intracellularly. Cell shape and size resembled *Histoplasma* and *Torulopsis*. The Gram smear was reviewed, and small, pinkish, oval budding yeast cells with halo around them were seen (*Image set 4.4*). This is characteristic of *Histoplasma*, since the structure appeared intracellular. The patient was HIV positive.

The fungal culture started to show mold on IMA within five days and two days later turned white mold, identified as *Histoplasma capsulatum* based on microscopic morphology seen under the microscope in lactophenol cotton blue preparation. The isolate was confirmed by DNA probe at the reference lab.

Small yeast cells are frequently encountered in the clinical specimens received by the microbiology lab, and their shape and size are so close and overlapping that only a few clues would aid in their exact structural identity. No matter how much experience one may have, *Histoplasma* present in the initial inoculum of the specimen in low number and without intracellular morphology would be hard to suspect. Care must be taken not to over-report direct smear results, since the clinician is going to act according to what has been reported on the patient's specimen.

Image set 4.4

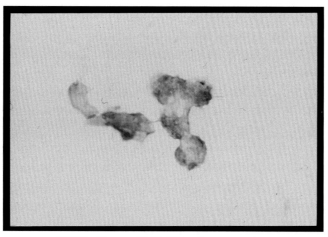

4.4 BAL: Gram x1000 *Histoplasma capsulatum*

4.5 Case 14

A mediastinal lymph node collected from a 58-year-old female was submitted to microbiology for C&S, TB, PCP, viral studies, and fungal culture. Gram results reported as "many pus seen, NBS." All microbiology tests other than fungal analysis requested were negative. A fungal stain showed many small oval budding yeast. The Gram smear was reviewed, and many small, oval yeast, narrow-base budding seen intracellularly as characteristic for *Histoplasma* (*Image set 4.5*). Fungal culture started to show mold on the IMA within four days; it was allowed to mature, followed by identification as *Histoplasma capsulatum* and confirmed by DNA studies done at the reference laboratory.

Image set 4.5

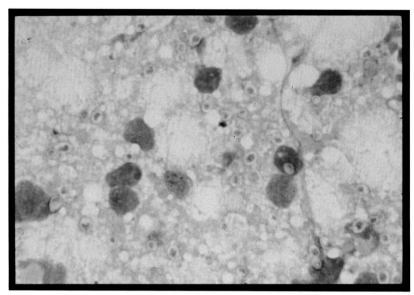

4.5 Lymph Node: Gram x1000 *Histoplasma capsulatum*

4.6 CASE 15

A BAL specimen from a 45-year-old female was received by microbiology for C&S, TB, PCP, viral studies, and fungal culture. The Gram smear results "no pus cells and NBS" were reported. All microbiology tests other than fungal analysis requested were negative. Very few small oval budding yeast cells (one structure seen was typical of *Histoplasma*) were seen in the fungal smear (CW). Pathology consulted microbiology for clarification about yeasts seen in fungal smear since pathology was not sure about its identity (*Image set 4.6*). *Histoplasma capsulatum* was recovered within nine days, identified, and confirmed by DNA probe.

The above three cases are unique in the sense that *Histoplasma* was suspected from the direct smear well before recovering the organism from the culture media. All three Gram smears were reviewed, and *Histoplasma*-like structures were observed in two of the three cases above. It is also interesting to know that pathology saw more structures on their smears but were unable to name it since the routine mycology smear could not really provide enough information to name an organism although it was thought to be *Histoplasma*. There are many such instances where pathology seeks mycology's help for interpreting organisms when they could not find mycology results on line since the physicians did not request certain clinical specimens for mycological analysis but other tests included histopathological investigation. The author's experience suggests that sometimes decisions made in histology by pathologists were corrected in mycology, since some irregularities that surfaced in the interpretation and results were either corrected or amended by the mycology expert. It would appear that the importance and the superiority of pathology smears remain very high over routine fungal-stained smears. As a result, pathology does not have much problem spotting the infected area and finding organisms easily due to the fact that their procedure produces high-quality direct smears maximizing specimen material on the microscopic slides. However, pathology does come across interpretive problems in defining structures, especially when fungal elements fail to appear similar to the textbook description.

When reading Gram smears (fungal smears or GMS), upon observing small oval budding yeast they must be compared with the size and shapes of similar-looking organisms, such as *Torulopsis* (*Candida glabrata*), *Malassezia*, and *Histoplasma*.

Sometimes low numbers of cells seen do not carry enough information to categorize small budding yeast into a distinct category due to the shape and size of the yeast and overlapping morphology of several other entities of the yeast group such as *Torulopsis*, *Histoplasma*, *Malassezia*, and other yeasts. In the above case, very few small oval budding yeast cells were seen. Not much information was available about the morphology of the structure seen except that a cell displayed morphology under the microscope suggesting *Histoplasma*. The Gram smear was reviewed, and very few oval cells, faintly stained, were seen. Due to the low number of cells seen, with insufficient information, it was decided not to report the structures as *Histoplasma*, but the information about suspicious *Histoplasma* seen was marked on the back of the work card hidden for laboratory use only.

A week later pathology consulted mycology after finding similar cells in the GMS smear and wanted our opinion for *Histoplasma*. GMS was reviewed by mycology, and the smear was found loaded with small oval budding cells consistent with *Histoplasma*.

Image set 4.6

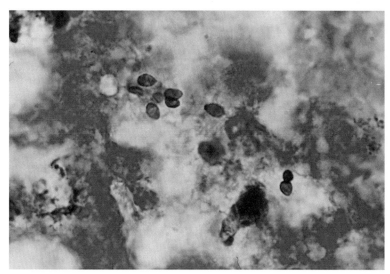

4.6 BAL: GMS x1000 small narrow-base budding yeast cells of *Histoplasma capsulatum*

4.7 CASE 16

A 58-year-old male with ALS, severely immunocompromised, was admitted to the hospital for systemic infection. He was on immunosuppression therapy as well as on antibiotic therapy. His blood culture was sent to the microbiology laboratory for C&S in automated blood-culture system known as BacT/Alert. On day three the blood-culture machine marked the bottle in the system as positive. The blood-culture bottle was pulled from the machine and processed for subbing on solid culture media. A Gram was also prepared from the positive blood-culture bottle. No bacteria were seen in the Gram smear, but yeast were seen and reported to the clinician. Culture plates were incubated at 37⁰C. Upon receiving a yeast report from the lab, the ward sent a new blood-culture set every day for the next few days to confirm yeast recovery.

After 24 hours of incubation, only thin precipitate was noticed on the surface of the blood agar and chocolate agar plate. All plates were re-incubated for another day. After 48 hours, no significant improvement in colonial morphology was seen, and the culture plates were forwarded to the department's mycology section where they were incubated at 28⁰C. Plates were examined after 48 hours, and a bluish-tinged mold was noticed covering the entire surfaces of blood and chocolate agar plates (*A: Image set 4.7*). At this point a plate contamination by *Penicillium* species acquired from the environment was assumed. A new subculture set was processed from the positive blood-culture bottle. The old set of plates was held in check.

Meanwhile, subsequent blood cultures received were positive for yeast that were later identified as *Candida kefyr* and *Candida krusei*. The new set of plates from the very first positive blood-culture bottle was examined after 48 hours incubation and showed the exact picture as before. No yeast was isolated from the first blood-culture set. Something interesting came to mind, and we questioned ourselves as to what happened to the yeast that was reported on the first day from the Gram smear made from the first set of the blood-culture bottle. This led mycology to review the Gram smear, and structures in filamentous form (septate hyphae) were found (B: *Image set 4.7*). The Gram smear provided enough information to suspect the isolate as *Histoplasma* and shipped the fungus immediately to the reference lab, where it was confirmed as *Histoplasma capsulatum* by DNA probe.

Those areas that led the Gram smear reader to suspect and report yeast to the ward in the first place were explored, and also why systemic fungus was not suspected by the IDS team knowing that the patient was not responding to the antimicrobial therapy. The Gram smears were re-examined from all positive blood-culture bottles, and some more interesting things were found in them. The very first set of blood culture had *Histoplasma* that went on to germinate in the blood-culture bottle, and the smear reader may have come across small oval pink cells in yeast morphology (C: *Image set 4.7*). The cells seen were small, oval, and displaying narrow-base budding resembling *Histoplasma*. Later on, when these cells began to germinate and formed septate hyphae in the blood-culture medium, they were not spotted by the smear reader. Upon review of the Gram smear, large amount of septate hyphae were seen, but very few pinkish-staining small oval-budding yeasts were seen in a few microscopic fields. The Gram smear from all the subsequent blood-culture sets showed budding yeast (D: *Image set 4.7*) and also grew yeast within 24 hours identified as *Candida kefyr* and *Candida krusei*. It was also noticed that all blood cultures growing yeasts also grew *Histoplasma* later on. However, the first culture growing *Histoplasma* did not grow any yeast.

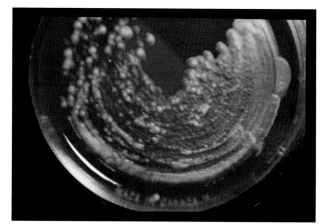

4.7-A Blood: bluish-green *Histoplasma capsulatum* yeast growing on BA

4.7-B Blood: Gram x1000 septate hyphae and small pink cells (*Histoplasma*)

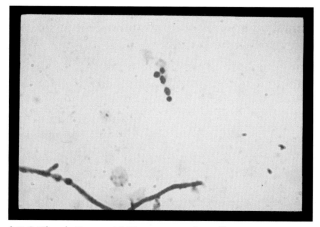

4.7 C Blood: Gram x1000 septate and small pink yeast (*Histoplasma*)

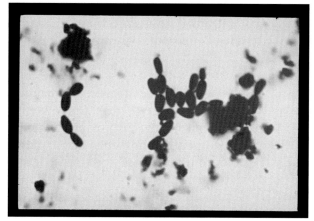

4.7-D Blood: Gram x1000 yeast (subsequent blood cultures grew *Candida kefyr*, *Candida krusei* and *Histoplasma*)

CHAPTER 5

5.1 COCCIDIOIDES[8, 9, 20, 23, 29, 30, 33, 42]

Coccidioides immitis is found in the desert southwestern United States, Mexico, and Central and South America. However, *Coccidioides* has been found in other non-endemic areas due to the ease of travel. The infection of coccidiomycosis is acquired by inhalation of infective arthroconidia of *Coccidioides immitis*. Infection initiates within the lungs and may resolve spontaneously and may progress and disseminate in immunocompromised hosts, infecting the skin, bone, brain, and other organs. The formation of spherule and the release of endospores cause granulomatous and suppurative reaction in the tissue.

Coccidioides immitis is a fast-growing fungus and is extremely hazardous. All work on *Coccidioides* must be carried out in level-three lab safety containment facilities using extreme safety precautions. Although *Coccidioides* grows as white, it may attain any color. In the lung, after rectangular arthroconidia (3–4 x 3–6 µm) have been inhaled they convert into spherules (10–80 µm) containing endospores (2–5 µm) that are released upon bursting open of the mature spherule. The recognition of *Coccidioides* in the direct smear prepared from the clinical specimen poses no problem when mature spherules filled with endospores are observed. Many times the textbook description of the fungi suspended in the colloidal material is not observed. As a result, the structures seen in the direct smear under the microscope are thoroughly studied in order to interpret them accurately.

Although Gram smear is not a classic staining procedure for the detection of dimorphic fungi, the author believes that justice has not been done by the microbiologists, mycologists, and histopathologists who profess the superiority of the specialized staining procedure to find fungi in the clinical specimens and discredit the most important first line of defense system such as the "Gram stain" to prove its usefulness. However, one must find a way to make Gram smear a helpful tool for finding fungi looking beyond bacteria. Histopathology does not always receive specimens for fungal investigation, and microbiology departments do not always receive clinical specimens for fungal investigation; specimens are processed only for C&S as requested even if fungi are present in them. During such instances, denying the patient care by not screening the Gram smear for the presence of fungi would be risky. Unreported fungi in direct smears would also delay the appropriate therapy to the patient. Such wrong decisions in not making use of the rapid and cost-effective procedure would underestimate the potential of the Gram stain method. Gram stain is not only an alternate to the diagnostic tool but has been proven useful by producing valuable patient-related diagnostic information that would go beyond the boundary of the bacteriology entity.

5.2 CASE 17

An open-lung biopsy specimen from a 64-year-old male was submitted to the microbiology lab for C&S, TB, viral, and fungal investigation. The patient had a nodule in the lung. The Gram smear was reported as "pus cells seen, NBS." The specimen was negative for TB and viral studies. The fungal smear stained by Fungi-Fluor (CW) showed occasional round, thick-walled cells, sitting side by side or still attached as if they were separating from each other. The microscopic size of the cells and morphology closely resembled *Blastomyces* (*A: Image set 5.2*). KOH prep showed very few single, round, and thick-walled cells (*B: Image set 5.2*)

The fungal smear results were reported out, and within the next 48 hours a white fluffy mold was isolated from blood agar (*C: Image set 5.2*) and chocolate agar (*D: Image set 5.2*). Culture plates (blood and chocolate agar) were opened in the bacteriology lab on the open bench during reading of the culture media. The fungus also grew on mycology culture media (IMA & BHIA) as white fluffy mold that did not appear to be *Blastomyces* by macroscopic morphology. It was unknown at this time that it was an extremely hazardous and level-three fungus that needed extreme precautions and biosafety level-three lab containment. Wet prep using lactophenol aniline blue was made and examined under the microscope. No conidia were seen. The isolate was subcultured on PDA and mycosel (medium containing cycloheximide). Growth occurred in PDA as white with pinkish tinge and spreading to glass; on mycosel the fungus remained fluffy white. Lactophenol aniline blue made from PDA now showed excellent barrel-shaped arthroconidia alternating with empty cell, identified as *Coccidioides immitis* (*E & F: Image set 5.2*), which was confirmed by the reference lab by using DNA probe. The recovery of this most hazardous fungus was a total surprise to us as well as to the clinician. We were expecting the fungus to be *Blastomyces*, since *Coccidioides* is not native in Canada. We enquired about the patient's traveling history and found out that the patient had been to Mexico, an endemic area for *Coccidioides*.

Image set 5.2

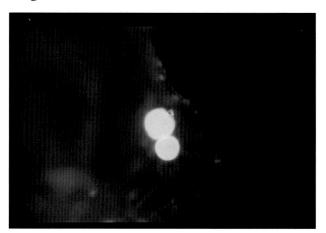

A BAL: CW x400 Immature spherules of *Coccidioides* resembling *Blastomyces* (note thickness of the cell wall is not clearly evident in CW)

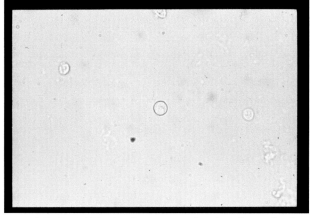

5.2-B BAL: KOH x400 Immature spherules of *Coccidioides immitis* resembling *Blastomyces* are clearly visible under the bright field (note thickness of the cell wall is clearly visible)

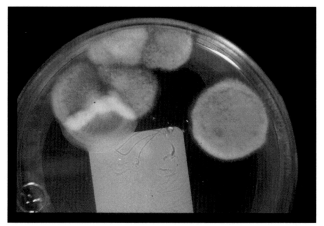

5.2-C BAL (Culture): *Coccidioides immitis* growing on IMA within few days (*Coccidioides*)

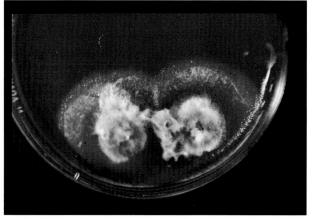

5.2-D BAL (Culture): *Coccidioides immitis* growing on BA rapidly becoming fluffy, spreading and turning mouse-grey

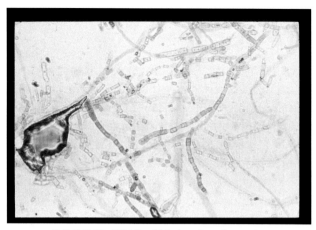

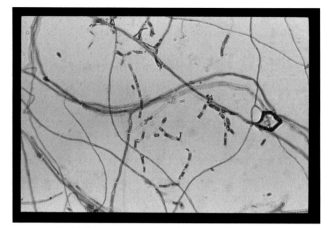

5.2-E BAL: LPAB x400 showing alternating
arthroconidia of *Coccidioides immitis*

5.2-F BAL: LPAF x400 showing alternating
arthroconidia of *Coccidioides immitis*

5.3 CASE 18

A BAL specimen from a 32-year-old female was received by microbiology for C&S, TB, viral, and fungal investigation. The patient had a variety of medical conditions and was initially diagnosed for TB after her TB skin test turned positive with 15 mm induration. Gram stain results showed no pus cells and NBS, and no fungal elements were seen in fungal smear. Two colonies of white mold grew on routine C&S culture medium (blood agar), and three colonies grew on inhibitory mold agar (IMA). Scotch prep (wet mount) was prepared using lactophenol aniline blue (LPAB) and examined under the microscope. Excellent microscopic structures such as alternating arthroconidia were observed, and the fungus was identified as *Coccidioides immitis* and confirmed at the reference lab using DNA probe. The fungus turned grayish and started to spread.

Upon recovering *Coccidioides* from the culture, the original smear was re-examined; only a few medium-sized non-budding round bodies were observed. The Gram smear was also reviewed and very few thick-walled round cells resembling immature spherules of *Coccidioides* seen.

This case is interesting for two reasons; first, the patient's symptoms were overlapping and the diagnosis for TB was made after a TB skin test turned positive. *Coccidioides* was not suspected but was a surprise for the physician when reported. Second, the Gram smear and the CW smear were both negative initially. After recovery of *Coccidioides* from the culture, both smears were reviewed and contained very few medium-sized round cells, suggesting immature spherules of *Coccidioides* (*Image set 5.3*). A low number of non-budding round cells would overlap morphology of similar-looking cells falling in different groups.

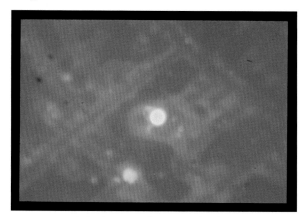

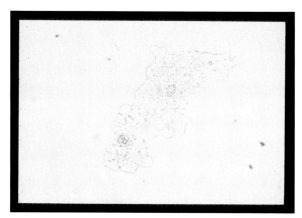

5.3-A BAL: CW x400 *Coccidioides immitis*
Observe thickness of the cell wall
mimicking double wall (immature spherule)

5.3-B BAL: WP x400 Immature spherules of *Coccidioides immitis* (CW smear was viewed under the bright field)

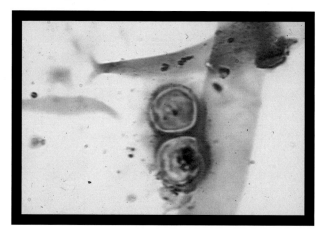

5.3-C BAL: Gram x1000 immature spherule of *Coccidioides immitis*
(immature spherules mimicking *Blastomyces*)

5.4 CASE 19

A 59-year-old male was admitted to the hospital for cough and shortness of breath and respiratory distress. A BAL specimen was sent to the lab for C&S, TB, PCP, viral, and fungal analysis. A Gram stain showed pus cells, but no bacteria were seen. All other tests were negative, except a white mold grew on C&S and fungal culture media within 48 hours. Lactophenol showed excellent alternating arthroconidia fitting into the morphology of *Coccidioides immitis,* which was confirmed by the reference lab using DNA probe. The Gram smear and the fungal smears were reviewed again. The Gram smear remained negative as before. However, the fungal smear showed few medium-sized, round, and thick-walled immature cells mimicking spherule (*Image set 5.4*).

Sometimes organisms are so low in number that it leads the smear reader to miss potential pathogen, as happened in this case. Even if rare structures are seen under the microscope, it would pose problems to accurately define the identity of the cells seen. Therefore, lab technologists must rely on the step-wise approach described in **Flowchart 2** to rule out other similar-looking structures based on minor details that separate one from the other. Patient data and the travel history also provide some groundwork to suspect potential pathogens such as *Coccidioides*. However, there is always a risk of missing fungal elements when the number of organisms remains low.

Had the travel history of the patient not been questioned, coccidioidomycosis would not have been considered to be a part of the diagnostic evaluation. There is lot to learn from the above cases. Symptoms common with other illnesses do not alert the clinician to ask for *Coccidioides* investigation as a differential diagnosis. People having mild symptoms after traveling to endemic areas should be screened for *Coccidioides*. Extreme safety precautions are necessary for the medical technologists during examining culture plates for bacterial identification to avoid accidental exposure to fungus growing on C&S culture media. Not all people inhaling conidia of *Coccidioides* develop the disease.[18, 20] The majority of people recover from the mild flu-like illness, suggesting that fungi do not invade human tissue unless the patient is immunocompromised.[24] Three main barriers— temperature, immunity, and nutritional requirements—tend to limit the establishment of fungi in humans before it is able to invade human tissue.[24]

When infection is non-progressive and begins to resolve spontaneously, the fungal element load tends to decrease. Therefore, a low number of organisms present in the specimen are most likely to be missed during direct microscopic examination of the specimen.

Image set 5.4

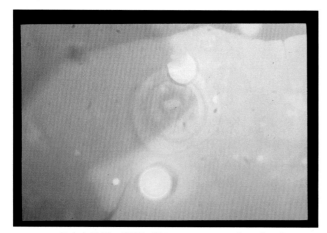

5.4-A BAL: CW x400 immature spherule of
Coccidioides immitis (one of the two appears opened or broken-up)

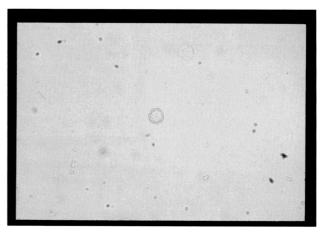

5.4-B BAL: KOH x400 immature spherule of
Coccidioides immitis

5.5 CASE 20

A STAT PCP analysis on a BAL specimen collected from a middle-aged South Asian male HIV-positive patient was received by the microbiology department for C&S and fungus. Direct smear revealed pus cells and normal flora in the Gram smear. The CW smear showed clusters of PC; several oval to round, non-budding cells in variable size (*A: Image set 5.5*) were seen as well. Few cells showed large, round, burst open from one side, and releasing endospores (*B: Image set 5.5*), characteristics of *Coccidioides*. Wet prep by KOH showed round bodies of medium size (*C: Image set 5.5*). The Gram was reviewed and a *Coccidioides*-like, mature, split open-and-empty spherule (*D: Image set 5.5*) was seen. Direct smear results were reported to the clinician. For comparison purposes the GMS smear showing a mature spherule, and endospores inside and outside the mature spherule are displayed (*E: Image set 5.5*). Within the next three days a white mold (*F: Image set 5.5*) spreading rapidly was recovered from culture and shipped to the reference laboratory, which confirmed *Coccidioides immitis* by morphological and DNA studies.

5.5-A BAL: CW x250 *Coccidioides immitis* - variable size of immature spherules (observe *Pneumocystis jiroveci cysts* in the background)

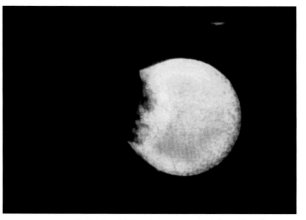

5.5-B BAL: CW x400 Mature spherule of *Coccidioides immitis* (observe faintly fluorescing small oval endospores packed inside the mature spherule)

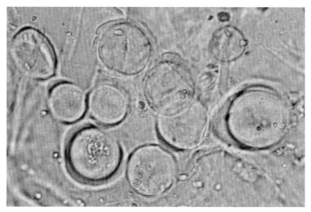

5.5-C KOH x400 *Coccidioides immitis* converted to spherule phase in converse medium at higher temperature (unrelated case)

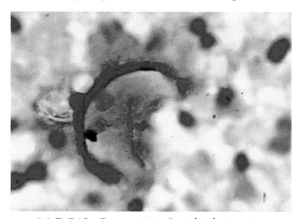

5.5-D BAL: Gram x1000 *Coccidioides immitis* (mature split open empty spherule without endospores)

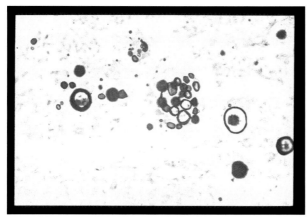

5.5-E Lymph Node: H & E and GMS x250 *Coccidioides immitis* - different stages

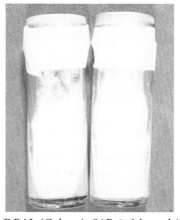

5.5-F BAL (Culture): SAB & Mycosel (medium containing cycloheximide) slopes side by side fast growing *Coccidioides immitis* rapidly filled the tubes

CHAPTER 6

6.1 *SPOROTHRIX*[8, 9, 20, 23, 29, 33, 42]

Sporothrix schenckii causes subcutaneous infections mostly associated with the lymphatic channel causing multiple sinus drainage. Initial infection may start as a small nonhealing ulcer on the finger or back of the hand, progressing to nodular lesions of the skin or subcutaneous tissue and becoming a chronic ulcerative infection. Sporotrichosis is commonly acquired by direct inoculation due to injury from plant material (rose thorn) or soil or by inhalation, causing pulmonary involvement. Sporotrichosis is a cosmopolitan disease found worldwide and a public health problem in certain rural areas of the world. Sporotrichosis may also disseminate and become fatal.

In direct clinical specimen, *Sporothrix* does not pose a problem in recognition. In Gram smear, it produces yeast cells of variable cell size ranging from oval, round, fusiform, and elongated (1–3 x 3–10 μm).[17] The most common structures, known as "cigar bodies," are often observed in direct smears made from clinical specimen. However, at times the yeast cells produced in the host tissue are inseparable from other yeasts and *Candida* species.

The fungus grows with a moist, pale, pasty, yeast-like consistency. It begins to change color, becoming tan, grey, and black with age. Microscopically, the mycelial phase produces two types of conidia; hyaline, teardrop to round conidia (2–3 x 3–6 μm), borne at the tip of slightly inflated conidiophore forming denticle rosette around it. The second type on conidial structure is known as "sleeve-shaped," dark brown conidia borne on the both sides of the hyphal stalk.[19]

6.2 CASE 21

Ear swab and skin biopsy specimens collected from alcoholic and immunocompetent hosts were processed for C&S and fungus. Gram stained smears showed similar structures in both that were easily identified as yeast, but the identity *Sporothrix schenckii* was confirmed only after recovering mold from the corresponding cultures. Smears were reviewed; yeast cells and "cigar-shaped" bodies were observed but were not considered different from other yeast cells by the routine microbiology technologist during Gram smear examination *(Image set 6.2)*.

Image set 6.2

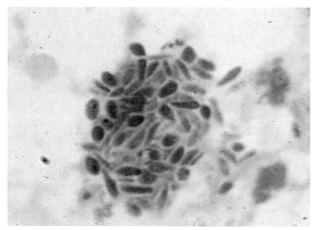

6.2-A Ear: Gram x1000 yeast cells *Sporothrix schenckii* (observe elongated yeast cells in cigar-shaped morphology)

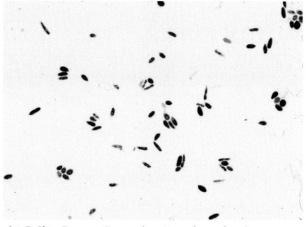

6.2-B Skin Biopsy: Gram x400 *Sporothrix schenckii* cigar-shaped in yeast phase.

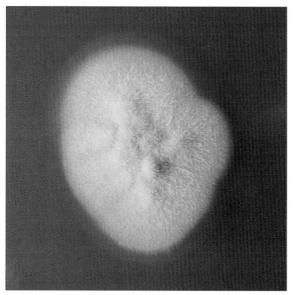

6.2-C Skin Biopsy (Culture): *Sporothrix schenckii* plate white, moist, yeast-like, low mycelium growing on PDA.

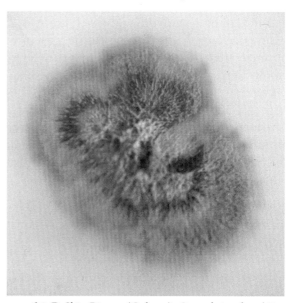

6.2-D Skin Biopsy (Culture): *Sporothrix schenckii* turning dark on SAB.

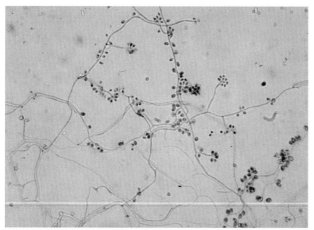

6.2-E LPAB x400 *Sporothrix schenckii* displaying rosette form morphology.

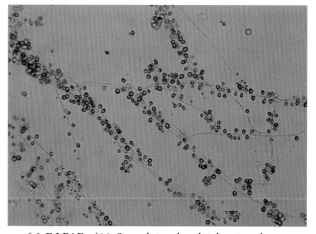

6.2-F LPAF x400 *Sporothrix schenckii* showing sleeve shaped structures.

CHAPTER 7

7.1 *PENICILLIUM MARNEFFEI*[8, 9, 20, 23, 29, 33, 42]

Penicillium marneffei is the only dimorphic species in the genus *Penicillium*. *P. marneffei* is a fungus causing disease of the reticuloendothelial system (RES). Small oval yeast-like cells (2.5–5 µm) are found in clusters within macrophages resembling *Histoplasma*. However, a closer look at the structures seen under the microscope would find that small cells of *P. marneffei* do not multiply by budding but by fission. A small septum is often seen dividing the cell into two. These cells continue to grow within the specimen and would change the morphology of the classic structure to larger form depending on the delay for processing the specimen after collection, especially if the collected clinical specimen has been left at the room temperature for too long. As a result, elongated forms like hyphae or arthroconidia may be seen in the direct smear specimen.

P. marneffei is most commonly found in Southeast Asia. Although the infection is more common among immunocompromised hosts, it also affects the immunocompetent hosts as well. The disease was first described affecting the RES of the Chinese bamboo rat that serves as a reservoir for this potential hazardous dimorphic fungus.

7.2 CASE 22

A 32-year-old male Vietnamese patient developed a non-productive cough, progressive dyspnea, fatigue, and skin rash. A BAL specimen was sent to the microbiology lab for C&S and fungal investigation with STAT request for PCP. Foamy exudate without the evidence of PCP was seen in the CW-stained smear. Initial Gram results were reported as NBS; however, upon review, foamy exudate and small, pink, oval, and non-budding cells were seen in the Gram-stained smear. One of the cells displayed a septum, indicating a bisected cell (*A: Image set 7.2*). In CW smear clusters of small (2–3 µm) bisected cells characteristic of *Penicillium marneffei* were seen (*B: Image set 7.2*). Upon re-checking the fungal smear (CW), *Pneumocystis jiroveci* cysts were observed. Another specimen smear fixed in alcohol was stained by GMS and showed excellent bisected cells characteristic of *P. marneffei* (*C: Image set 7.2*). The patient's blood culture also turned positive and showed septate hyphae (*D: Image set 7.2*), indicating dissemination. The fungus grew fast on C&S and mycological media as blue-grayish-green (*E: Image set 7.2*) and produced diffusible red pigment in the medium that appeared more intense in reverse side of the plate (*F: Image set 7.2*). LPAB revealed *Penicillium* structures, and the fungus was identified as *P. marneffei,* which was later converted to yeast phase at 37°C. *P. marneffei* is a rapid-growing dimorphic fungus and a potential pathogen infecting the reticuloendothelial system (RES). *P. marneffei* usually infects immunodeficient patients. It may also cause infection in an immunocompetent host. *P. marneffei* is endemic in Southeast Asia, where the bamboo rat serves as a reservoir for this deadly disease.

Several important issues surround this case as well as the safety concern for healthcare workers dealing with a fungus that has the ability to infect immunocompetent hosts. The patient was suffering from AIDS and developed pneumonia. The clinician suspected PCP, obtained BAL, and sent STAT for PCP studies along with other microbial analysis. CW (Fungi-Fluor) done on the specimen failed to pick up cysts of *Pneumocystis* due to the greater degree of foamy exudate present in the specimen suppressing PC cysts. However, bisected cells of *P. marneffei* that were present in the smear were not picked up by the smear reader until after the smear was reviewed. PCP was positive by IFA testing. The Gram smear that was originally reported as NBS was correct from the bacterial point of view. Upon review, few oval cells in small-yeast morphology were observed in the Gram smear mimicking *Histoplasma* but was ruled out after bisected cells were observed under the microscope.

The CW smear was reviewed, and this time interesting structures were seen. Within foamy exudate, some cysts of *Pneumocystis* were found in single and in small groups. Excellent small, oval, bisected cells were seen in clusters depicting the characteristic morphology of *Penicillium marneffei*. C&S and mycology cultures were processed two days prior. C&S and mycology culture were examined immediately and were found with bluish, greenish, and grayish mold with intense red pigment diffused in the medium (best seen in IMA and EBM plates). C&S plates were opened on the bench during bacterial examination. The bench technologist noticed mold growing on the plate and transferred the culture to the mycology section without knowing that the extremely hazardous fungus was exposed in the laboratory environment. The fungus was easily identified and also converted at 37⁰C, showing fragmentation and producing bisected cells. *Penicillium marneffei* is a dimorphic fungus residing in the reticuloendothelial system (RES). Its intracellular morphology frequently resembles *Histoplasma* when observed in direct smears stained by GMS and other methods.

Image set 7.2

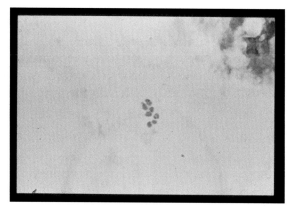

7.2-A BAL: Gram x1000 *Penicillium marneffei* (often confuse with *Histoplasma*. Observe bisected cell with septum)

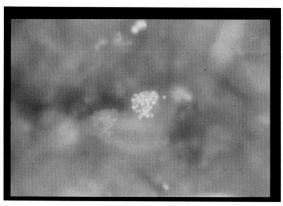

7.2-B BAL: CW x400 *Penicillium marneffei* (Observe bisected cell with septum)

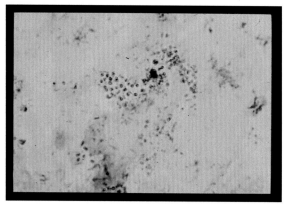

7.2-C BAL: GMS x1000 *Penicillium marneffei* (excellent clusters of bisected cells)

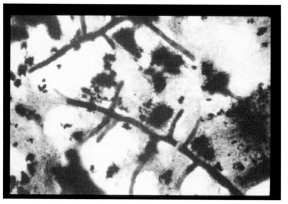

7.2-D Blood: Gram x1000 *Penicillium marneffei* started to covert into filamentous forms showing arthroconidia-like structures in blood culture bottles at 370C in BacT/Alert automated blood culture system.

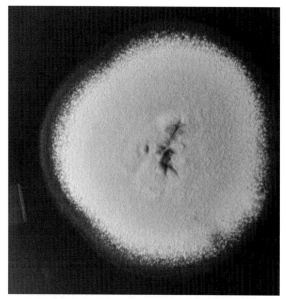

7.2-E BAL (Culture): Diffusible red pigment produced on PDA by *Penicillium marneffei* (surface)

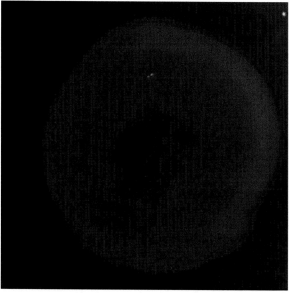

7.2-F BAL (Culture): Diffusible red pigment produced on PDA by *Penicillium marneffei* (reverse).

CHAPTER 8

8.1 OPPORTUNISTIC FUNGI (COMMON HYALOHYPHOMYCOTIC AGENTS) [8, 9, 20, 23, 26, 27, 29, 33, 42]

Generally, all fungi are opportunistic (author's experience). However, some opportunistic fungi are able to cause fungal infections among human hosts who are less protective against fungal infections. The opportunistic category of fungi in this section belongs to the fungi that are hyaline or lightly colored. *Aspergillus* species in this genus are frequently isolated from clinical specimens but not necessarily implicated with disease process. However, the isolation of *Aspergillus fumigatus* in this genus is more virulent. Therefore *Aspergillus fumigatus* isolated from the clinical specimen collected from an immunocompromised host must be considered significant. Other opportunistic fungi that are recovered from the clinical specimens such as *Fusarium*, *Acremonium*, *Scedosporium*, *Paecilomyces*, *Trichoderma*, and *Scopulariopsis* are less commonly isolated as compared to *Aspergilli*. *Penicillium* is the most common environmental contaminant and most frequently isolated from clinical specimens. However, its recovery does not pose any threat to patient care. Since most *Penicillium* species are blue green, macroscopic morphology is similar to *Aspergillus fumigatus*. Therefore its recovery and reporting is usually indicated as "normally non-pathogenic."

Recovery of fungi from clinical specimens is not difficult. However, the recovery of fungi is always matched with the clinical setting so that an appropriate decision may be made to require antifungal therapy in a timely fashion. Initiation of the majority of fungal infections involves inhalation of fungal spores or fragments from air. Individuals whose immunity is lowered due to immunosuppressive medication or severe debilitating disease often acquire an invasive pulmonary or sinus infection, from where it may spread to the surrounding tissue or disseminate to other organs. Fungemia is most commonly seen with *Fusarium*. Noninvasive fungal infections are found in debilitated patients as well as in immunocompetent hosts with a preformed lung cavity or having a lung cyst.

Most fungi are able to grow at lower temperature (< 30⁰C) and are therefore unable to grow at human body temperature. The most opportunistic fungal infection caused by *Aspergillus fumigatus* being thermo-tolerant is not only able to grow at 37°C but also grows well above that (> 50⁰C). *Aspergillus fumigatus* is ubiquitous in nature and frequently inhaled from the environment. Conidia (2.5–3.5 μm) of *A. fumigatus* inhaled by immunocompetent hosts reach alveolar spaces and are readily cleared by alveolar macrophages.[26, 27] Conidial germination that escapes macrophage surveillance is arrested by circulating neutrophils (PMNs) initiating phagocytosis of conidia and fungal hyphae (3–12 μm wide).[26, 27] *Aspergillus fumigatus* causes infections with a wide range of spectrum such as benign colonization, allergy, or invasive aspergillosis. However, the situation is different among immunocompromised hosts, since inhaled conidia begin to germinate to cause invasive infection.[26, 27]

Therefore, the recovery of fungus in clinical specimen must be correlated with the corresponding direct smear results. Opportunistic fungi, when seen in direct microscopic smears (Gram, KOH, or CW), may indicate the clinical significance of the prevalent fungus; it is at least justified to say upon observing fungal elements in the direct smear that the recovery of corresponding fungal species from the culture media is real. In direct smears, septate hyphae are 3–12 μm wide. Septate hyphae more or less remain parallel, a useful clue for separating aseptate hyphae having irregular width and being too wide in some areas and too narrow at the other end.

The dark pigmented fungi (dematiaceous), Zygomycetes, and other fungi causing opportunistic infections are discussed separately.

8.2 CASE 23 – *ASPERGILLUS FUMIGATUS*

A female in her 80s suffered from chronic sinusitis. Maxillary sinus contents were submitted to the microbiology lab for C&S and fungal investigation. The Gram smear was reported as "occasional pus cells and no bacteria seen." The fungal smear (CW) showed many septate hyphae in mycology (*A: Image set 8.2*). The Gram smear was reviewed and septate hyphae were found beyond recognition. They appeared hidden due to the specimen debris making the detection difficult (*B: Image set 8.2*). In order to make these fungal elements stand out, the Gram smear was overstained by CW/KOH (*C & D: Image set 8.2*). As expected, blue green mold was isolated in C&S and mycology culture media within 48 hours (*E: Image set 8.2*). Lactophenol wet prep was examined under the microscope. Uniseriate arrangement of the phialides, covering two-thirds of the vesicle and having smooth-walled conidiophore, identified the fungus as *Aspergillus fumigatus* (**F:** *Image set 8.2*).

This was a simple and straightforward case but was not picked up by the Gram smear reader because fungi are not searched for in the Gram-stained smears on a routine basis. Fungal elements masked by the specimen debris were an added disadvantage for its detection. The focus to find bacteria remains the prime object in clinical microbiology, and the Gram smear reader would make little effort to specifically search areas for clues leading to spot organisms other than bacteria. The importance of observing organisms other than bacteria in the Gram stained smear adds up to the quality of work done by the microbiology staff. Missing other organisms, such as fungi, in the Gram smear would in turn delay the process of appropriate therapy for the patient.

Image set 8.2

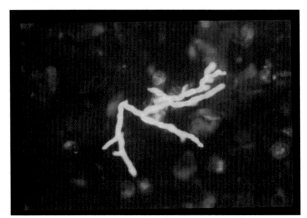

8.2-A Max. Sinus: CW x400 septate hyphae
(*Aspergillus fumigatus*)

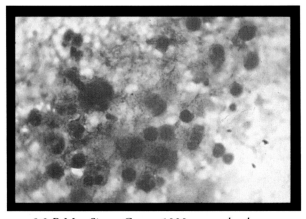

8.2-B Max Sinus: Gram x1000 septate hyphae
(hidden) (*Aspergillus fumigatus*)

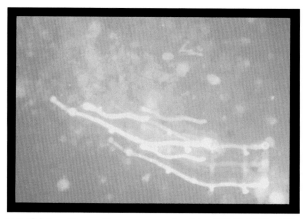

8.2-C Max. Sinus: Gram over-stained by KOH-CW x400 septate hyphae (clearly observable under the UV light)

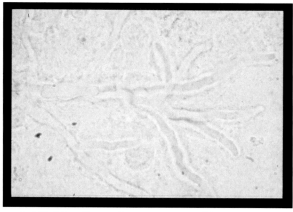

8.2-D Max. Sinus: Gram over-stained by KOH-CW x1000 Septate hyphae (clearly observable under the bright field bright-field)

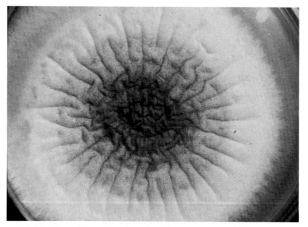

8.2-E Max. Sinus (Culture): Blue-green mold on IMA (*Aspergillus fumigatus*)

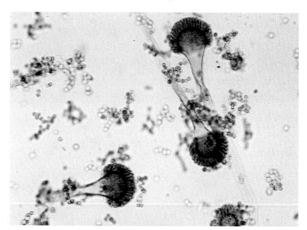

8.2-F Max. Sinus: LPAB x400 uniseriate phialides covering 2/3 of the vesicle (*Aspergillus fumigatus*)

8.3 CASE 24 – *ASPERGILLUS FLAVUS*

A 55-year-old male transplant patient had fungal sinusitis. Left maxillary sinus tissue was sent for fungal analysis. The pathologist examined histology-stained smears (GMS) and found wide fungal hyphae and reported as *Mucor* (*A: Image set 8.3*). The clinician consulted mycology, asking for clarification since mycology reported as septate hyphae on the same specimen. Histology smears were reviewed, and abundant septate hyphae in wider width were seen. The pathologist was notified, and an amended report was reissued from pathology. Gram stain from the same specimen sent to microbiology encountered no problem, and septate hyphae (*B: Image set 8.3*) was immediately reported to the clinician. The culture grew *Aspergillus flavus* (*C & D: Image set 8.3*).

Aspergillus flavus usually produces septate hyphae with acute angle (dichotomous) branches like they do in other *Aspergillus* species. However, *Aspergillus flavus* has been seen producing wider septate hyphae resembling Zygomycetes. Histology smears are usually processed in a way that the specimen is held in an embedded wax block. Preparation of the smear involves cutting wax blocks in thin sections so that a single cell layer of the specimen is applied to the microscopic glass slide. During cutting waxed specimens in thin layers, fungi are often cut in different ways based on the angle at which they are cut. As a result, filamentous fungi display different morphological structures under the microscope such as hollow tubes, holes, or fungal elements without septation. Fungi present in the clinical specimens and processed by histological techniques are frequently cut at the site of the fungal filament just above or below the septum in the fungal hyphae. This would lead to misidentification of fungi with wider hyphae appearing as non-septate, especially when they are seen in shorter length without septation. One must always keep a key point in mind to avoid making incorrect interpretation; that Zygomycetes are wider hyphae but have a variable width of the cell wall. The smear reader must screen the entire slide to look for clues when wide hyphae do not show any septum in certain microscopic fields but may have more structures to demonstrate irregular width since the internal pressure in long non-septate hyphae fluctuates, making aseptate hyphae appear too wide in some areas and too narrow in the adjacent vicinity.

Pathology has several specialized procedures to perform direct microscopy on clinical specimens, producing excellent smears stained by GMS & PAS for the detection of fungi. The clarity and the quality of the histology smears are always superior to the routine staining procedures such as Gram, KOH, and CW done in the microbiology laboratory. Pathology staff are able to detect fungal elements without a problem. However, pathology often finds difficulty in interpreting structures seen in GMS, PAS, or H&E smears. Therefore, close contact for consultation between pathology and microbiology is extremely useful in order to correctly identify microscopic structures seen in direct smears made from clinical specimens.

The patient discussed above is a lung-transplant recipient. Therefore, this individual is made immuno-compromised on purpose. Any fungal species may gain access to an immunocompromised host, causing infection leading to invasiveness. Among filamentous fungi, *Aspergillus* species and Zygomycetes both have been recovered from the clinical specimens from patients who fall into this group. However, the identity of the fungus involved is extremely important for the proper management and the initiation of prompt antifungal therapy to the patient. For Zygomycetes and *Aspergillus* species, the clinician may select a different set of antifungal agents for treating the patient.

Images set 8.3

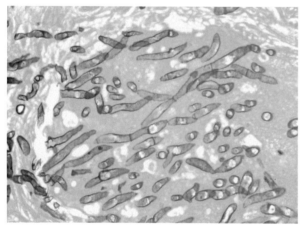

8.3-A Max. Sinus: GMS x400 septate hyphae
(pathology reported as *Mucor*)

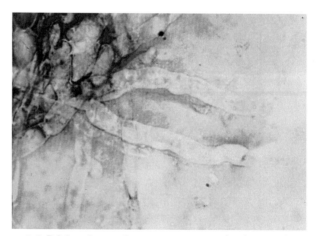

8.3-B Max. Sinus: Gram x1000 septate hyphae (wide hyphae may have confused the pathologist)

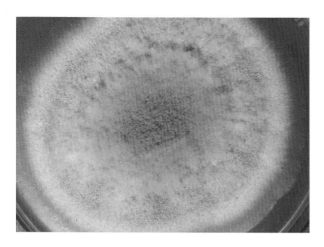

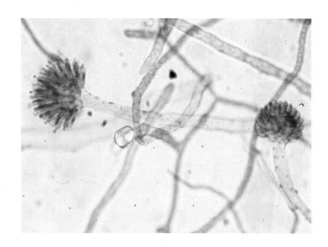

8.3-C Max. Sinus (Culture): Yellow-green mold growing on IMA (*Aspergillus flavus*)

8.3-D Max. Sinus: LPAB x400 biseriate & uniseriate radiating phialides, rough-walled conidiophores (*Aspergillus flavus*)

8.4 CASE 25 – NASAL BIOPSY

This is another example of incorrect interpretation of fungal elements seen in sinus biopsy specimens and reported as "*Mucor* present" by pathology. The same sinus biopsy specimen was also processed in microbiology for C&S and fungal analysis. CW smears done in mycology showed septate hyphae, and the results were reported. The clinician immediately called mycology asking for clarification; as a result the direct smear done in mycology as well as GMS & PAS smears done in the pathology department were reviewed. The patient was a post-lung-transplant recipient, and they wanted to know if *Mucor* interpretation by pathology was correct.

Histology special-stained smears are superior to the routine microbiology staining procedures. The pathologist would not usually miss fungal elements in specimens stained by specialized procedures. However, categorizing fungi seen under the microscope in histology smears may be difficult. This is an example of a straightforward case; the smears clearly showed septation in the fungal hyphae present in the GMS & PAS smears but were interpreted incorrectly. The structures present in the histology smears do not indicate any confusion, and it is not clear why it was called *Mucor* without supporting evidence. The most important clue in this smear is the septation followed by unique width in septate hyphae. Aseptate hyphae are wider but also have irregular width in them. Septate hyphae more or less display parallel sides of the hyphal fragment with frequent septation. Aseptate hyphae, on the other hand, are wider, like ribbon. However, the width of the aseptate hyphae is variable, with rare or no septum. Rare septum in the aseptate hyphae is a necessity since the cell has to close at some point. The distance of the septum is greater, and as a result it does not always show in a microscopic field. Cellular contents within the large cell fluctuate between the two ends during physical and mechanical pressure; as a result the walls of the aseptate hyphae become much wider at one end and very narrow at the opposite end, having lesser pressure of the cytoplasmic stream. The pressure and other factors applied during the specimen processing result in damaging the cells to the point where they become non-viable; as a result many times aseptate hyphae are seen in the specimen but the culture remains sterile.

Aspergillus flavus is known to display much wider hyphae, causing confusion with Zygomycetes. Therefore, it is suggested that the frequency of septation and the parallel walls of the fungal hyphae must be observed before ruling out aseptate hyphae. Gram smears from this specimen showing much wider hyphae were interpreted as septate hyphae because the walls of the fungal hyphae are parallel and frequent septation is observed (*A & B: Image set 8.4*). Frequent septation and dichotomous branching, if seen; provide the best clues for the identification of the fungal hyphae as septate hyphae.

Image set 8.4

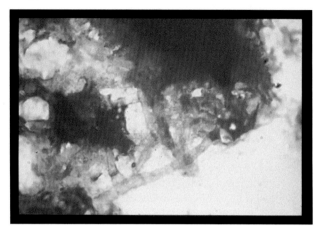

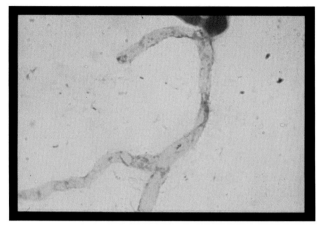

| 8.4-A Max Sinus: Gram x1000 septate hyphae (*Aspergillus flavus*) | 8.4-B Max Sinus: Gram x1000 septate hyphae (*Aspergillus flavus*) |

8.5 CASE 26 – *ASPERGILLUS NIGER*

A 70-year-old lung-transplant patient developed respiratory illness and was admitted to the hospital. BAL was received by the microbiology lab for C&S and fungal investigation. The fungal smear stained by CW showed fungal elements mimicking yeast/pseudohyphae. Fungal conidia were also observed in other microscopic fields (*A: Image set 8.5*). The Gram smear results were reported as "pus cells and commensal flora seen," without the mention of fungal elements. Upon review, the Gram stain showed fungal hyphae and also fungal conidia. No yeast or pseudohyphae were seen (*B: Image set 8.5*). Culture grew dark mold (*C: Image set 8.5*), identified as *Aspergillus niger* (*D: Image set 8.5*) on both C&S and mycology culture media.

Aspergillus fumigatus is the most common fungus recovered from respiratory specimens collected from immunocompromised hosts. *Aspergillus niger* and flavus have also been isolated from the clinical specimens collected from this type of patient population. Direct smears made from clinical specimens usually show septate hyphae of dichotomous nature. However, fungal conidia may be seen easily in direct smears while septate hyphae may be hidden or remain undetected for other reasons. Care must be taken not to call fungal conidia yeast; therefore, always observe the presence or absence of budding that is the most readily available clue for dividing yeast and fungal conidia. Another valuable morphology of the cell in question is to carefully observe the pink-staining cytoplasm extending around the boundary of the non-budding fungal conidia.

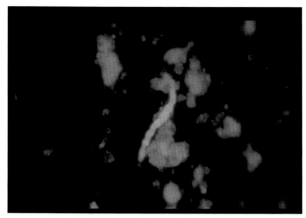

8.5-A BAL: CW x400 fungal elements (*Aspergillus niger*)

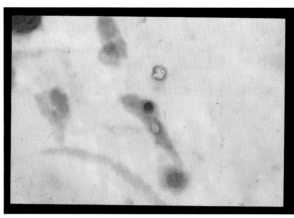

8.5-B BAL: Gram x1000 fungal elements (*Aspergillus niger*)

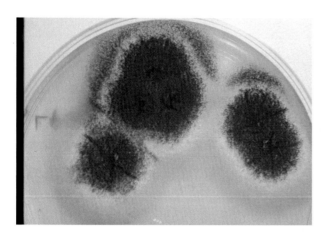

8.5-C BAL (Culture): Jet black mold on IMA (*Aspergillus niger*)

8.5-D LPAB x400 biseriate phialides
covering entire vesicle
(*Aspergillus niger*)

8.6 CASE 27 – *ASPERGILLUS TERREUS*

A 23-year-old immunocompromised, double-lung-transplant female was admitted to the hospital with high fever and respiratory distress. Her condition started to deteriorate, and she was kept in intensive care. A BAL specimen was sent to microbiology for C&S and fungal analysis. PMN was seen in the Gram smear, but no bacteria were seen. The fungal smear in mycology showed fungal elements that appeared septate and also appeared slightly pale (*A: Image 8.6*). Wet prep of the specimen smear showed slightly darker fungal elements with septation (*B: Image set 8.6*). The Gram smear was reviewed, and septate hyphae were seen that firmly stained the interior of the fungal elements with marked cell wall, mimicking dematiaceous fungus (*C: Image 8.6*). At this point, nothing was consistent to confirm structures as dematiaceous fungus, and it became an urgent issue to get a specialized test, Fontana Masson (melanin stain), done on the specimen. Melanin stain (FM) turned out to be negative (*D: Image 8.6*). Within the next three days, a pale, granular, with low mycelium fungus appeared on the culture plates and slowly turned beige/tan and cinnamon colored (*E: Image 8.6*). A lactophenol prep showed structures consistent with *Aspergillus terreus* (*F: Image 8.6*). It is important to distinguish hyaline fungal elements from the dematiaceous group, because the clinician might select an appropriate antifungal agent to treat the patient. Fungal infections caused by the dematiaceous group are known to resolve slowly. Fungal infections caused by *Aspergillus terreus,* on the other hand, may show resistance to amphotericin B. Once again we re-examined our smears to learn what made us suspicious about dark fungus and not hyaline. Hyaline fungi such as *Aspergillus fumigatus* are the most opportunistic pathogens for immunocompromised patients and are often seen in direct smear specimens. They appear as parallel-walled hyaline hyphae with frequent septation, and dichotomous structures are usually observed in direct smears under the microscope. In the above example, the structures appeared to be like "toruloid" and darkened areas inside the cell, which confused our judgment. Care must be taken to define structures correctly, since the interpretation of the direct smear would have an immediate impact on the patient care. We were happy that we did not issue results prematurely that would have required corrective action later on.

Image set 8.6

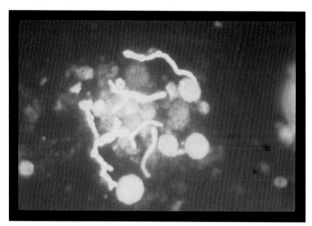

8.6-A BAL: CW x400 fungal elements, sinuous and showing divisions in fungal elements i.e. septate hyphae

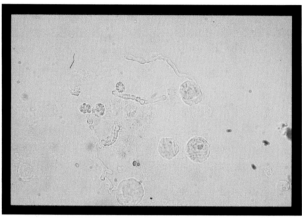

8.6-B BAL: KOH x400 fungal elements showing septum is more evident along with dichotomous at end; however, pale looking tinge in fungal hyphae required to rule out pigmented hyphae.

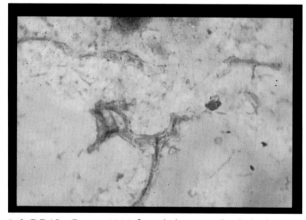

8.6-C BAL: Gram x1000 fungal elements clearly looking as septate hyphae. Pale-reddish brown fungal cell wall and foot cell type structures are also noticed

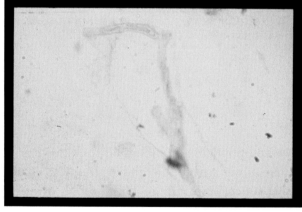

8.6-D BAL: FM x1000 non-pigmented fungal elements are observed except the interior of the cell marked deeply like a foot cell. Possibility of dematiaceous fungus is ruled out

8.6-E BAL Culture: light brown to cinnamon colored mold on *Aspergillus terreus* on SAB

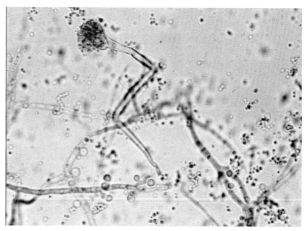

8.6-F LPAB x400: biseriate phialides cover top 2/3 of the vesicle and round spherical aleurioconidia (microconidia) of *Aspergillus terreus* are observed

8.7 CASE 28 – *ACREMONIUM RECIFEI*

A 49-year-old post-lung-transplant male patient went in the woods and got injured by a tree branch falling on his leg. Ten days later, a lump began to appear at the site of injury. The patient went to the local hospital and was treated with antibiotics without resolution. The patient was referred to the IDS team. Leg aspirate was drawn from the infected site and sent for C&S and fungal culture. Initially the Gram stain results were reported as "pus cells seen, no bacteria seen." Fungal smear showed fungal elements not consistent with *Aspergillus* (*A: Image 8.7*). The Gram smear was reviewed, and the fungal elements that had previously been missed were observed and appeared as true septate hyphae (*B & C: Image set 8.7*). Mycology culture media grew a glabrous type of fungus on IMA and other media used (*D: Image set 8.7*). Quick wet mount using LPAB was made, and structures seen under the microscope were interpreted and reported out to the clinician as *Fusarium* species (*E: Image set 8.7*) pending speciation from the reference laboratory. The specimen was shipped to the reference laboratory. However, the slide culture was also set up. A few days later our slide culture reached maturity and a wet prep using LPAB was made; structures seen under the microscope were in direct conflict with the previously reported *Fusarium* species. The slide culture showed structures resembling *Acremonium* species (*F: Image 8.7*). Meanwhile the reference laboratory analyzed the isolate, calling it *Verticillium* species and added a qualifying statement of "normally non-pathogenic." We did not accept the identification for two reasons: 1) The fungal elements were present in the direct smear specimen, indicating the clinical significance; 2) the patient was post–lung transplant and therefore must be immunosuppressed on purpose. The isolate was submitted to the mycology experts (another reference lab), who did complete analysis on the isolate and correctly identified the fungus as *Acremonium recifei*. The microbiology technologists analyzing fungal species must pay attention to the clinical data of the patient, the site, and the nature of fungal infection in order to make an appropriate reporting statement. The patient died due to unexplained causes.

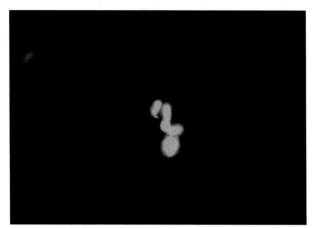

8.7-A Thigh aspirate: CW x400 fungal elements seen

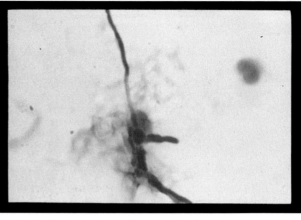

8.7-B Thigh Aspirate: Gram x1000 fungal elements in septate morphology

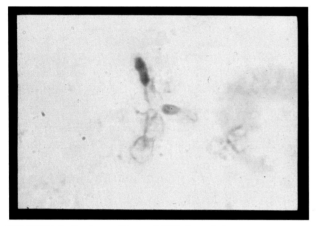

8.7-C Thigh Aspirate: Gram x1000 fungal elements not looking like yeast/ pseudohyphae

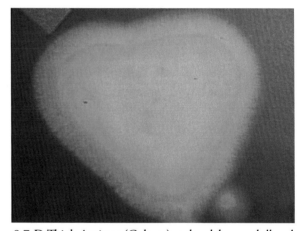

8.7-D Thigh Aspirate (Culture): pale, glabrous, dull and moist mold on IMA

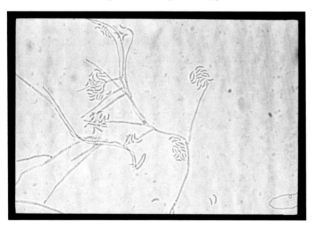

8.7-E Thigh Aspirate: LPAB x400 small curved conidia were first suspected of *Fusarium* species later identified as *Acremonium recifei* by the reference lab

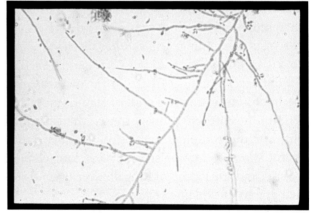

8.7-F Thigh Aspirate: LPAB x400 erect phialides producing small conidia at the tip identified as *Acremonium recifei* by the reference lab

8.8 Case 29 – *Fusarium solani*

Ethmoid contents from a 21-year-old female patient were submitted to the microbiology department for C&S and fungal investigation. The Gram smear results revealed pus cells, Gram positive cocci, yeast, and pseudohyphae. The fungal smear results were also reported as yeast with pseudohyphae until Gram and CW were reviewed after recovering a beige, pinkish, and slightly grey mold (*C: Image set 8.8*) isolated from all C&S & fungal culture media. The fungal elements seen in CW (*A: Image set 8.8*) and Gram (*B: Image set 8.8*) smears had no yeast but septate hyphae, which confused the smear reader due to the presence of swellings in the hyphal fragment. There were no constrictions at the joining points, and septation was clearly observed except that the hyphae did not display acute-angle (dichotomous) branching. Mold isolated from the corresponding culture was identified as *Fusarium solani* (*D: Image set 8.8*). Direct smears made from the clinical specimens do not display similar structures (any one particular fungal species) in every specimen they are recovered from. However, minor details like frequent septation along the length of the fungal hyphae, no budding forms, or pseudohyphae are sufficient to report structures as septate hyphae even if no dichotomous forms are observed. A true yeast must be searched for budding forms, and pseudohyphae must be distinguished by their specific structures such as constrictions between the two elongated yeast cells remaining attached. A closer look at the fungal hyphae would distinguish pseudohyphae being centrally swollen as compared to parallel walls of the true septate hyphae.

Image 8.8

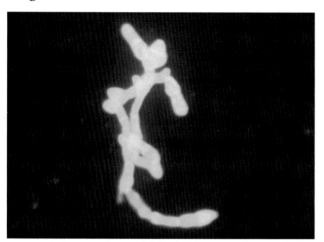

8.8-A Ethmoid Biopsy: CW x400 Septate hyphae (*Fusarium*)

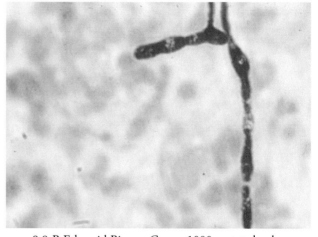

8.8-B Ethmoid Biopsy: Gram x1000 septate hyphae

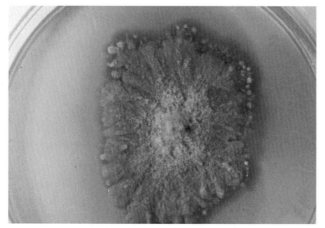

8.8-C Ethmoid Biopsy (Culture): beige, greyish mold growing on SAB

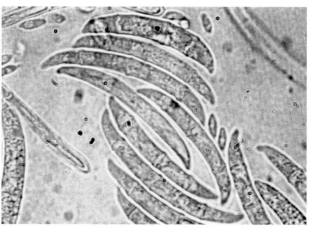

8.8-D Ethmoid Biopsy : LPAB x1000 curved, canoe-shaped multiseptate macroconidia of *Fusarium solani*

8.9 CASE 30 – *FUSARIUM OXYSPORUM*

A swab from the scrotum of a 64-year-old patient having redness and itching was received in microbiology requesting C&S and fungal culture. The nature of the specimen was such that no fungal culture was set up on swabs collected from the skin surfaces except that a Gram stain or calcofluor stain was done to report the presence or absence of yeast as per departmental policy. No pus cells and no yeast were seen in the Gram smear. A CW smear in mycology displayed fungal elements as pseudohyphae and septate hyphae (*A: Image set 8.9*). No yeast was seen in the entire smear. The Gram smear was reviewed and showed similar structures to those seen in the CW smear (*B: Image set 8.9*). At this point, culture was set up to find out about the fungus seen in the direct smear, producing structures resembling pseudohyphae without the evidence of yeast. Within a few days a pale white woolly mold (*C: Image 8.9*) appeared on the fungal culture media and was identified as *Fusarium oxysporum* (*D: Image 8.9*). A review of the original Gram and CW smears found structures as phialides and/or conidia of *Fusarium* that had previously been thought of as pseudohyphae. It is interesting to know that fungi do not produce conidia in tissue. However, the nature of the specimen being superficial skin swab, fungi were able to grow in an area with lower temperature (< 35⁰C), which allowed the fungus to invade superficial skin and produce filamentous forms, phialides, and hyphae. There are no inhibitory factors present on the surface of skin to prevent the fungus from producing phialides and conidia that clearly show in the Gram image (*C: Image set 8.9*). When a fungus is present in a deep-seated area of the human body, it usually does not produce conidia.

Image set 8.9

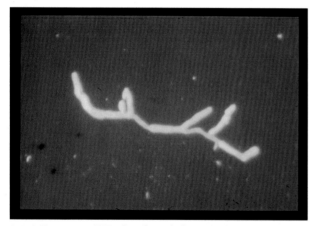

8.9-A Scrotum: CW x400 fungal elements (appears to be in septate morphology)

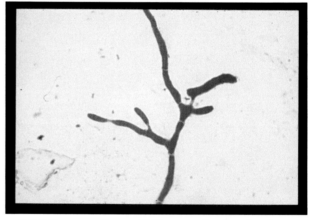

8.9-B Scrotum: Gram x1000 septate hyphae showing structures resembling phialides and macroconidia of *Fusarium* species

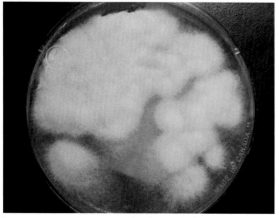

Image set 8.9-C Scrotum (Culture): pale-white woolly mold on IMA (*Fusarium* species)

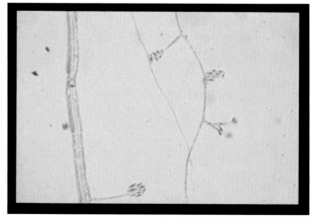

8.9-D Scrotum: LPAB x400 curved conidia of *Fusarium oxysporum*

8.10 CASE 31 – *SCEDOSPORIUM APIOSPERMUM*

A 34-year-old right-corneal-transplant recipient male patient presented with severe corneal abscess, perforation, and decreased vision after being splashed by garbage-contaminated fluid. Symptoms progressed from a mild sensation to intense ocular pain. He visited the local hospital where he received antibiotics from the outpatient department. His condition continued to worsen, and he was admitted in the acute care teaching hospital. Corneal scraping was sent to microbiology for C&S and fungus. No fungus was seen in direct Gram and fungal smears. Repeated corneal scrapings were done. Bacterial cultures yielded no growth until a single colony of *Scedosporium apiospermum* was isolated from the fungal culture. Corneal infection due to *S. apiospermum* is rare. The fungus was reported to the clinician by phone, and results were also posted on line. The isolate was also sent to the reference lab for antifungal-susceptibility testing, since *Scedosporium apiospermum* shows resistance to most antifungal agents. The Gram smear was overstained by KOH/CW, and septate hyphae that were not picked up earlier from the Gram-stained smear were now seen (*Image set 8.10*). Repeated attempts were made to induce the isolate to sexual form, but it failed to convert into the sexual stage.

Image set 8.10

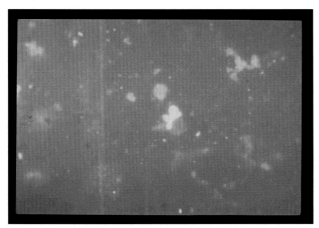

8.10-A Cornea: CW x400 fungal elements (undifferentiated)

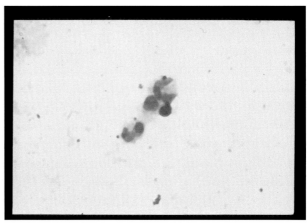

8.10-B Cornea: Gram x1000 fungal elements not clearly seen (Gram was later over stained by KOH-CW procedure; image unavailable)

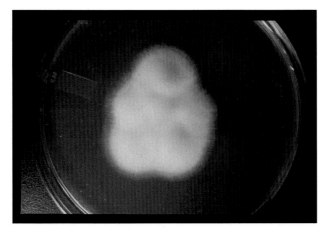

8.10-C Cornea (Culture): light colored (beginning to turn greyish) woolly mold on IMA

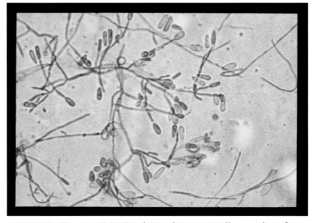

8.10-D Cornea: LPAB x400 solitary annelloconidia of *Scedosporium apiospermum* borne on short or elongated conidiophores

8.10-E Cornea: LPAB x400 asexual synnemata (Graphium) of *Scedosporium apiospermum*.

8.11 CASE 32 – *TRICHODERMA LONGIBRACHEATUM*[43]

A male patient was suffering from chronic sinusitis. The patient had a history of atopy and asthma. A sinus biopsy was received by microbiology for culture and susceptibility and fungal analysis. Direct Gram smear results were reported as yeast (*A & C: Image set 8.11*) seen. A fungal stain (CW) showed fungal elements with hyphal morphology (*B: Image set 8.11*). Within the next three days, a rapid-growing fungus was isolated from the culture (*D: Image set 8.11*) and identified as *Trichoderma longibrachiatum*[29] (*E: Image set 8.11*). No yeast was isolated. After a week, a fluffy, grayish, and woolly fungus appeared at the periphery of the *Trichoderma* colonies. A LPAB showed a long, thin, conidiophores-like structure with terminal oval-shaped conidia structures resembling *Scedosporium* (*F: Image set 8.11*). However, these structures also appeared inside the hyphal stalk as intercalary swellings, suggesting chlamydoconidia of *Trichoderma*.

Upon review of the Gram stain, the reported yeast cells were recognized as chlamydospores of *Trichoderma*. The patient was successfully treated with a combined management of sinus lavage, oral corticosteroids, itraconazole, and allergen immunotherapy. As in the previous case, a more thorough initial evaluation of the Gram stain would have led to prompt, appropriate therapy (*Image set 8.11*).

Image set 8.11

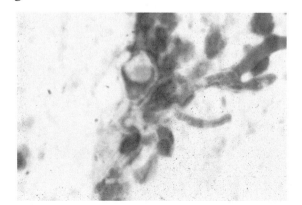

8.11-A Sinus Biopsy: Gram x1000 fungal elements appear true hyphae (initially reported as yeast)

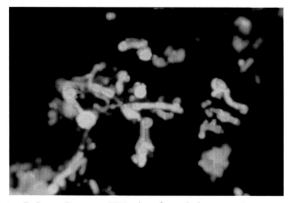

8.11-B Sinus Biopsy: CW x400 fungal elements (appears to be septate hyphae) reported as yeast

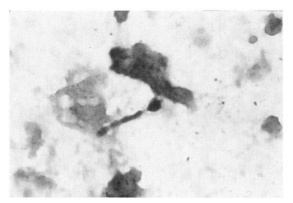

8.11-C Sinus Biopsy: Gram x1000 fungal elements appear true hyphae (initially reported as yeast)

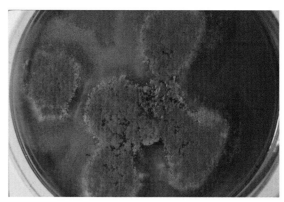

8.11-D Sinus Biopsy (Culture): Littman agar growing fragile green mold surrounded by hyaline cloudy and spreading second type of mold

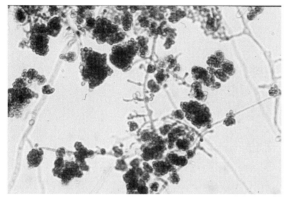

8.11-E Sinus Biopsy: LPAB x400 right angle inflated bottle-shaped phialides on branched conidiophores bearing round or ellipsoidal conidia of *Trichoderma longibracheatum*

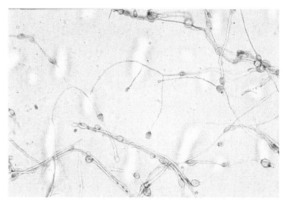

8.11-F Sinus Biopsy: LPAB x400 round, oval structures (resembling *Scedosporium*) are internal swellings or chlamydoconidia of *Trichoderma longibracheatum*

CHAPTER 9

9.1 DEMATIACEOUS FUNGI[6, 8, 9, 10, 20, 23, 29, 33, 42]

Fungi producing olive-grey, brown, or black (melanin) pigment in the cell wall of the fungal hyphae or conidia are classified as dematiaceous (dark) fungi. Most fungi in dematiaceous group produce grey to dark color from the very beginning as they first appear on culture media. However, some dark fungi may start off as white and pale at first, turning darker upon further incubation.

Fungi producing conidia enclosed within brown to dark-colored round bodies such as pycnidia, cleistothecium, or perithecium are not included in the dematiaceous group.[28]

In the clinical microbiology laboratory, simple tests such as Gram stain and wet preparation may be very useful to detect dark fungi during direct microscopic examination of the clinical specimen. Many times the shape of the fungal elements showing pale to dark boundary of the cell wall may indicate the presence of a dark fungus in the clinical specimen. Not all dematiaceous fungi would show a pigmented cell wall in a Gram smear or wet prep made from clinical specimens. However, suspicion of dark fungus may also be considered from the structures dematiaceous fungi produce in the tissue observed under the microscope in Gram, CW, or KOH wet prep. Dark fungi display structures unique to their category, such as cells in chains mimicking pseudohyphae or frequent swellings (toruloid) along the length of a hyphal fragment. A closer look at the joining ends would indicate the septation and not constrictions. It is important to demonstrate melanin pigment in suspected fungi, and this must be confirmed by a specific melanin stain such as Fontana Masson. Early detection of a phaeohyphomycotic agent present in the clinical specimen may provide relevant information to the clinician to select appropriate antifungal therapy for the patient.

9.2 CASE 33 – NONSPORULATING FUNGUS

A 52-year-old immunocompromised male patient was admitted to the hospital with a deep-seated nodule on the foot. Abscess aspirate was sent to the lab for C&S and fungal analysis. The Gram smear results reported as pus cells seen; no bacteria seen. The C&S culture was negative for bacteria. A fungal stain showed a moderate amount of fungal elements that appeared to be dematiaceous (*A: Image set 9.2*). The Gram smear was reviewed and displayed fungal structures not fitting regularly occurring septate hyphae morphology but more like dark fungus (*B: Image set 9.2*). The fungal smear (CW) was viewed under the bright-field microscope, and dark pigmentation on the cell wall were observed (*C: Image set 9.2*). Pigmented hyphal walls were confirmed by FM stain (*D: Image set 9.2*). About two weeks later a dark fungus (*E: Image set 9.2*) was recovered that remained non-sporulating to date (*F: Image set 9.2*).

This case is interesting since distorted-looking fungal elements were present in the direct Gram smear but were not picked up by the smear reader. Upon review, the fungal elements were evident in the Gram smear. However, the appearance of the structure was not noticeable enough to alert the smear reader about fungus. In the Gram image (B), the filamentous form is twisted and does not show straight parallel-walled hyphal fragments that are centrally bulged and appear like a swing hanging in the air. The Gram is even more interesting since the structures depict morphology of the dark fungus. In order to confirm melanin pigment in the cell wall, the smear was stained by melanin stain (Fontana Masson), which turned positive (slightly melanin pigment observed). The degree of intensity of dark pigmentation depends on the amount of melanin synthesized. Upon making a decision about pigmented fungal elements seen in the direct smear, the direct smear results were updated as "dematiaceous fungi" present. This information is necessary for the clinician to select an appropriate antifungal agent for therapy. Two weeks later dark fungus was isolated that is still pending for identification in the reference laboratory.

Image set 9.2

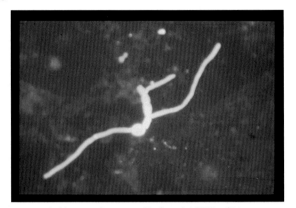

9.2-A Abscess Aspirate (Lt knee): CW x400 septate and toruloid hyphae

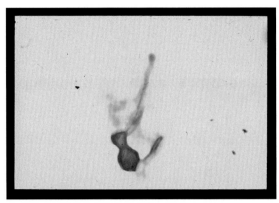

9.2-B Abscess Aspirate (Lt knee): Gram x1000 septate and toruloid hyphae

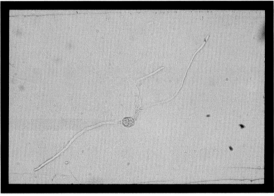

9.2-C Abscess Aspirate (Lt knee): KOH x400 septate hyphae

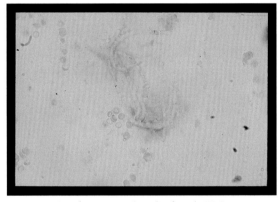

9.2-D Abscess Aspirate (Lt knee): FM x1000 septate hyphae some areas clearly showing brown pigment including several septa staining brown.

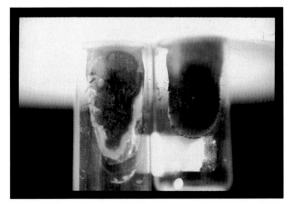

9.2-E Abscess Aspirate (Lt knee-culture): growing dark mold on PDA

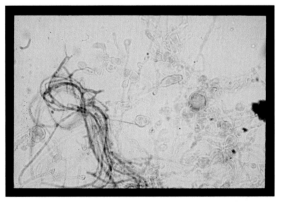

9.2-F Abscess Aspirate (Lt knee): LPAB x400 unidentified and non-sporulating fungus of dematiaceous group

9.3 Case 34 – Foot lump sent for C&S only

A skin swab specimen from a lump on the left foot of a male patient in his early 40s was received by the microbiology laboratory for culture and susceptibility only. No fungal culture was requested. The Gram stain showed pus cells, but no bacteria were seen. However, light brown pigmented and septate hyphae were observed in the Gram upon review (*A: Image set 9.3*). Wet prep was also done, and pale pigmented fungal elements were observed (*B: Image set 9.3*). This prompted the inclusion of a fungal culture workup, although it was not originally requested by the clinician. The Gram stain results were updated, indicating the presence of dematiaceous fungus. The clinician was notified, and immediate action was taken to alter the antifungal therapy to cover for phaeohyphomycosis. The patient began to respond to the antifungal therapy well before the final identification of *Exophiala jeanselmei* (*C & D: Image set 9.3*) was made three weeks later.

Image set 9.3

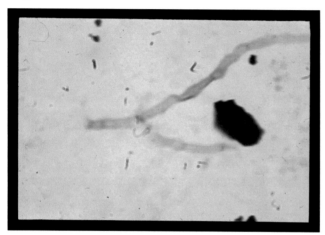

9.3- A Lump on foot: Gram x1000 dark septate hyphae

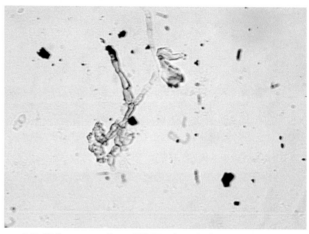

9.3- B Lump on foot: KOH x400 dark Septate hyphae.

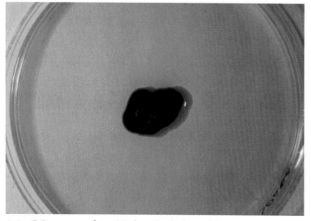

9.3- C Lump on foot (Culture): dark mold growing on SAB (*Exophiala jeanselmei*)

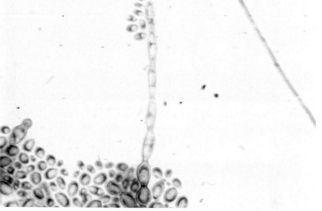

9.3- D Lump on foot: LPAB x1000 oval hyaline or pale brown and mostly unicellular conidia held in slime at the tip of annellide identified as *Exophiala jeanselmei*

9.4 Case 35 – Contents maxillary sinus

A 21-year-old female had chronic sinus. Maxillary sinus contents were sent to microbiology for C&S and fungal investigation. The Gram smear results were reported as pus cells, mixed bacteria, and yeast with pseudohyphae seen. The fungal smear stained by CW revealed septate hyphae of two different kinds: 1) hyaline; and 2) dark-walled mimicking dematiaceous group (*A: Image set 9.4*). Wet prep was examined under the bright-field microscope and pigmented cells were observed (*B: Image set 9.4*). The Gram stain results were corrected, and an amended report was issued. It was noticed from the Gram smear that hyaline and dematiaceous fungal elements (*C & D: Image set 9.4*) were clearly observed, and no special stain was needed to demonstrate melanin pigment in the cell wall of this particular dematiaceous fungus. The culture grew *Fusarium* species, and *Exophiala dermatitidis*.

Image set 9.4

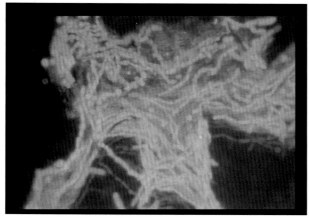

9.4-A Max Sinus Biopsy: CW x400 septate hyphae
(2 types of septate hyphae were observed)

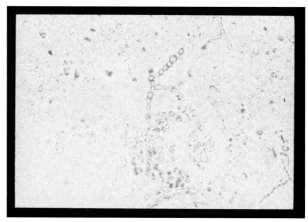

9.4-B Max Sinus Biopsy: KOH x400 septate hyphae
(hyaline and pigmented septate hyphae are seen)

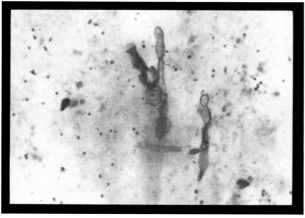

9.4-C Max Sinus Biopsy: Gram x1000 septate hyphae
(pigmented) *Exophiala dermatitidis* was isolated from culture

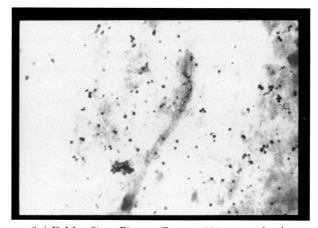

9.4-D Max Sinus Biopsy: Gram x1000 septate hyphae
(non-pigmented). Culture also grew *Fusarium* species

9.5 CASE 36 – TWO TYPES OF YEAST-LIKE FUNGI (*C. LUSITANIAE* & *EXOPHIALA DERMATITIDIS*)

A 32-year-old male lung-transplant patient was admitted to the hospital. He developed pneumonia-like symptoms, and a lung biopsy specimen was collected and sent to microbiology for C&S and fungal investigation. The Gram smear results were reported as 1+ PMN and 2+ yeast seen. The fungal smear (CW) in mycology also showed cells of yeast morphology (*A: Image set 9.5*). Culture grew yeast within 48 hours, identified and reported as *Candida lusitaniae*. Upon further incubation of the media, a second type of mucoid black colonies began to appear after 10 days that was identified as *Exophiala dermatitidis* (*B: Image set 9.5*). Upon isolating a second type of black yeast, the Gram smear and fungal smear were reviewed and demonstrated two types of fungal elements of yeast and mold morphology (*C: Image set 9.5*). Since fungal smear stained by CW would not display dark cell wall under the UV light, the CW smear was reviewed under the bright field, and darkness on the cell walls of certain yeast and filamentous form were noticed (*D: Image set 9.5*). Sometimes Gram and wet preparation are suitable to demonstrate pigmented fungal elements. However, in the event that no pigment is observed, the smear must be stained by the special stain such as FM to detect melanin pigment in the cell wall of dematiaceous fungi. Direct smear showed distinctly two types of structures: 1) clearly hyaline yeasts and pseudohyphae; and 2) structures distinctly displaying brown pigment in the walls of hyphal fragments and also showing toruloid type of structures often seen in dematiaceous group. *Candida lusitaniae* (*E: Images set 9.5*) & *Exophiala dermatitidis* (*F: Image set 9.5*) were recovered from the culture.

Image set 9.5

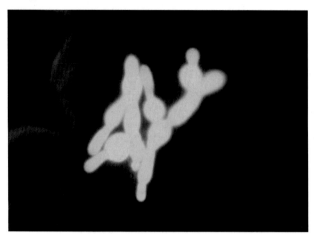

9.5- A Lung Biopsy: CW x400 yeast-pseudohyphae like (toruloid structures do not display pigment)

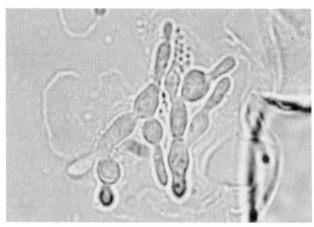

9.5- B Lung Biopsy: WP x400 yeast-pseudohyphae like dark pigment clearly visible)

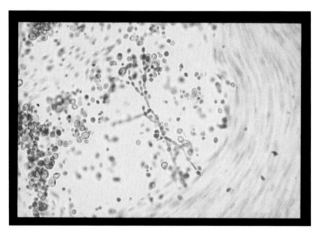

9.5- C Lung Biopsy: Gram x1000 yeast and pseudohyphae like (2 types structures seen)

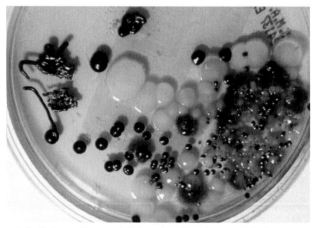

9.5- D Lung Biopsy (Culture): pale-white and dark moist yeast like two types colonies growing on IMA

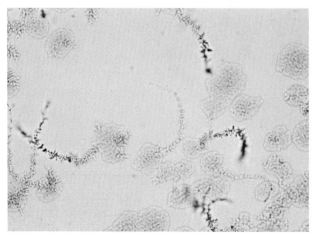

9.5- E Cornmeal Agar: x250 *Candida lusitaniae* showing short and curved pseudohyphae

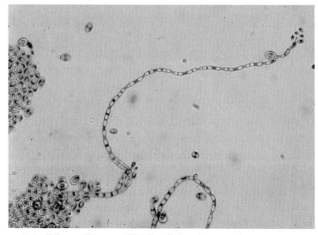

9.5- F Lung Biopsy: LPAB x400 microscopic structures identified as *Exophiala dermatitidis*

9.6 CASE 37 – *EXOPHIALA* SPECIES

An abscess specimen from a male patient was received by microbiology for C&S and fungal investigation. The Gram smear results were reported as "PMNs seen; no bacteria seen." In mycology, a CW smear showed interesting structures that appeared as yeast and pseudohyphae morphology (*A: Image set 9.6*). Since no yeasts in budding form were seen in the entire microscopic field, the UV light filter was switched to bright filter to observe pigmentation that is not observable in CW (fluorescent) smear. Wet prep showed structures that fit into the category of dark fungi such as swellings and pseudohyphae-like darker fungal elements (*B: Image set 9.6*). It was decided at this point to see what was present in the Gram smear that may have been missed by the smear reader. The Gram smear was reviewed, and caterpillar-like structures resembling pseudohyphae were seen without any individual yeast cells in budding form (*C: Image set 9.6*). The structures seen in the Gram were slightly tinged with dark pink pigment. A preliminary report of the presence of dark fungus was reported to the clinician. A fresh smear from the specimen was prepared and stained by the special melanin (FM) stain, and fungal cell walls displayed melanin pigment (*D: Image set 9.6*). About a week later poor growth on mycology media began to appear that turned dark upon maturity (*E: Image set 9.6*) after a few days. Slide culture was set up to capture the fungus in natural state. LPAB prep was made from the slide culture, which showed annellidic conidiation of *Exophiala* species (*F: Image set 9.6*).

There are other fungi in the dematiaceous group that have been observed in the clinical specimens collected from immunocompromised hosts causing local and systemic infections. The trick that is most useful to apply during direct examination of the clinical specimens is to observe hyphal fragments as they stand out clearly. What follows next is to define structures as they appear. Dark fungi (dematiaceous) usually do not show septate hyphae in dichotomous fashion as is often displayed in hyaline fungi. Dark fungi usually appear as cells in chains with prominent swellings along hyphal fragment. Such structures often resemble pseudohyphae. At this stage, the smear reader must search for yeast cells (single or in budding forms) present in other microscopic fields. For this reason, restrict your interpretation and do not report yeast or pseudohyphae unless a true budding form has been demonstrated at some point in the direct Gram or other fungal smear. When dark fungi are observed in the Gram smear, a wet prep using KOH should be made to observe dark pigmentation in the cell walls of fungal elements. Only if the smear reader is convinced about the dark fungi seen in the direct smear should the clinician be notified. Dark fungi grow slowly; therefore the direct smear results may be extremely useful to suspect phaeohyphomycotic agent (dark fungus).

Image set 9.6

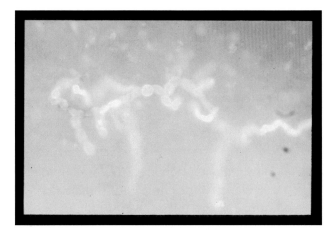

9.6-A Abscess Aspirate: CW x400 septate hyphae (toruloid)

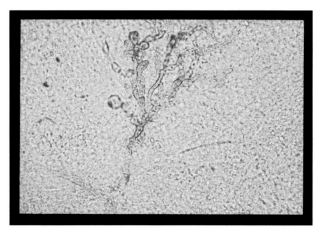

9.6-B Abscess Aspirate: WP x400 pigmented septate hyphae (toruloid).

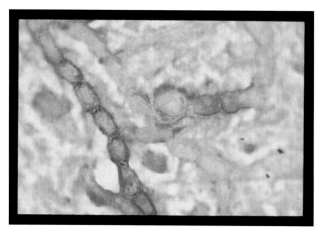

9.6-C Abscess Aspirate: Gram x1000 septate hyphae displaying pigment (toruloid)

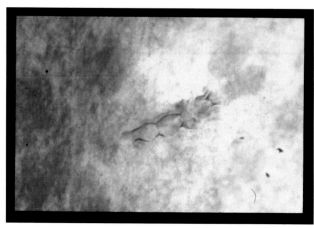

9.6-D Abscess Aspirate: FM x1000 pigmented septate hyphae (toruloid & dark)

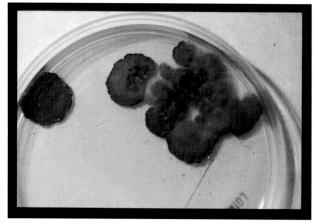

9.6-E Abscess Aspirate (Culture): dark mold growing on IMA

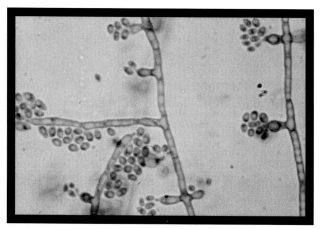

9.6-F Abscess Aspirate: LPAB x400 conidiophores hardly differentiated from vegetative hyphae, structures identified as *Exophiala* species.

9.7 Case 38 – *Alternaria*

A 40-year-old female had a sore on her foot that did not go away using disinfectants or antiseptic cream or by oral and local antibiotic therapy. A swab from the foot was submitted to microbiology for C&S and fungal investigation. Gram results were reported as "PMN seen; no bacteria seen." A fungal smear stained by CW showed a moderate amount of septate hyphae (*A: Image set 9.7*). The original Gram smear was retrieved and reviewed. Thin septate hyphae with irregular swellings (toruloid structures) along hyphal fragment were observed (*B: Image set 9.7*). There was no indication of melanin pigment in the cell wall of fungal hyphae seen in the Gram-stained smear or the wet prep. About a week later a dark fungus (*C: Image set 9.7*) appeared on mycology culture plates identified as *Alternaria* species (*D: Image set 9.7*).

The fungal hyphae seen in Gram and CW smears did not display any pigmentation in the cell wall of septate hyphae; however, long, thin, nondichotomous structures with irregular swellings (toruloid) should alert the smear reader to suspect dark fungus from the corresponding culture. The specimen smear should have been stained by FM to demonstrate melanin pigment in the cell wall.

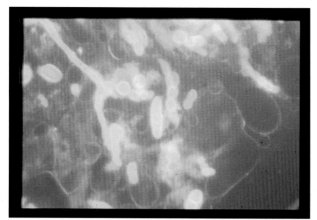

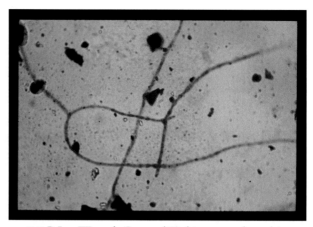

9.7-A Leg Wound: CW x400 fungal elements mimicking pseudohyphae morphology, no clear budding forms are seen

9.7-B Leg Wound: Gram x400 thin septate branching hyphae, no dichotomous or other details are seen

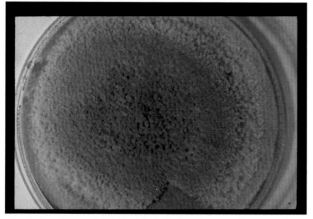

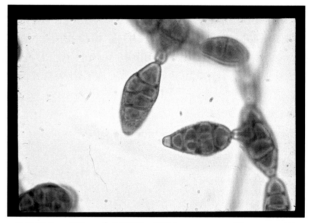

9.7-C Leg Wound (Culture): light brown, woolly, granular mold growing on SAB

9.7-D Leg Wound: LPAB x400 brown, muriform, ovoid poroconidia showing chain and apical beakile structure of *Alternaria* are observed

9.8 CASE 39 – *PHAEOACREMONIUM*

Synovial fluid from the knee collected from a 54-year-old male patient was received by the microbiology laboratory for C&S only. The Gram stain showed pus cells, and no bacteria were seen. The blood agar plate started to show growth of tiny colonies, pitting in the agar (*A: Image 9.8*). A Gram smear of the colonies revealed thin branching bacillary forms, and the culture was thought to be *Nocardia* (*B: Image 9.8*). Modified kinyoun (MK) stain (regular ZN stain was not included) was performed on the isolate and interpreted as positive for partial acid fast (*C: Image set 9.8*). *Nocardia* isolated was reported to the clinician pending confirmation from the departmental mycology section. All smears were reviewed in the mycology section. The isolate was also subcultured on specific selective and differential medium such as sodium pyruvate agar (PYRA), which did not agree with the isolate fitting in actinomycete entity. The original Gram smear did not show any organisms but hollow unstained tracks suggesting fungal elements (*D: Image 9.8*). The Gram stain made from the culture showed branching and septation with the width of hyphae >1µm (*B: Image set 9.8*), suggesting the involvement of a fungus. Therefore the culture disagreed with the *Nocardia* identification. Modified kinyoun was positive for partial acid fastness due to the fact that fungi do retain carbol fuchsin, the chief ingredient in the reagents of MK & ZN used for demonstrating acid fastness in mycobacteria and *Nocardia*. The ZN was not included along with MKS.

At this point, the culture was suspected of having a fungus that was a bit tan in color on primary culture, turning grayish. Further work continued to resolve the conflict. Culture smears were made and sent to the histology lab for GMS and melanin stain (FM). The specimen was also cultured on mycology media (IMA), which slowly grew as grayish and turned dark with age. The original Gram smear was processed to overstain with KOH/ CW stain (WP), which revealed the presence of slightly pale septate hyphae, indicating dark fungal elements (*E: Image set 9.8*). The earlier report sent to the ward was corrected, sending an updated report of "fungus isolated." GMS stain confirmed the fungus, and melanin pigment in the cell wall was confirmed by FM stain. The culture on initial C&S media started to turn dark after a week. Dark mold was also noticed on mycology media such as IMA and SAB. A slide culture set up identified the fungus as *Phaeoacremonium*, confirmed by the reference lab as *Phaeoacremonium inflatipes* (*F: Image set 9.8*).

The failure of the Gram stain to pick up any organism is due to the colloidal material that masked fungal elements, making Gram reagents unable to reach the target site. Upon overstaining the original Gram smear using KOH/ CW, septate hyphae were clearly seen. What confused the technologist reading C&S culture plates to suspect *Nocardia* was the undergrown young fungal colonies growing on the routine microbiological media behaving like aerobic actinomycetes since the isolate on C&S media was gritty and difficult to move. When modified kinyoun stain was performed, it gave a positive partial acid fastness result due to the fact that the fungal cell wall allowed basic fuchsin to enter and remain in the interior of the fungal cells during the decolorization step using dilute sulfuric acid. As a result, thin branching fungal elements were considered a bacterial entity such as *Nocardia*, which was reported but needed corrective action later on.

Image set 9.8

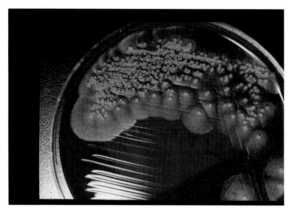

9.8-A Knee Aspirate (Culture): Blood Agar growing granular, gritty colonies were suspected of *Nocardia*

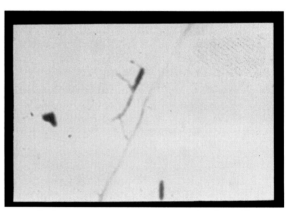

9.8-B Knee Aspirate: Gram x1000 thin branching pattern of filaments staining pink were initially reported as *Nocardia*

9.8-C Knee Aspirate: Modified Kinyoun x1000 showing partial acid-fast reaction, structures seen were initially reported as *Nocardia*

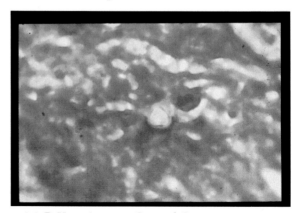

9.8-D Knee Aspirate: Original Gram x1000 upon review did not show *Nocardia* structures. Hollow unstained areas and fungal elements (see central area)

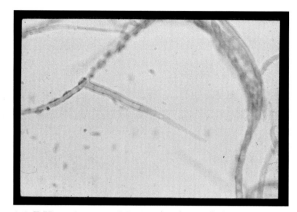

9.8-E Knee Aspirate: KOH x400 original Gram smear over stained by KOH/CW fungal elements are clearly seen

9.8-F Knee Aspirate: LPAB x400 long, dark phialides with narrow tip bearing small oval conidia were confirmed as *Phaeoacremonium inflatipes* by the reference lab

9.9 CASE 40 – *RAMICHLORIDIUM MACKENZEIE*

A 60-year-old male born in Somalia and suffering from hepatitis C was admitted to the hospital with confusion. A computerized tomography (CT) scan showed a left temporal parietal cystic mass. The initial diagnosis was pyogenic abscess, tuberculoma, or brain tumor. A brain biopsy specimen was submitted to microbiology for C&S and fungal investigation. The Gram stain was initially reported as pseudohyphae, but later review found that the structures were septate hyphae (*A: Image set 9.9*). The next day, the IDS team rushed to mycology looking for answers about the septate hyphae that were seen in the direct Gram smear. Their major suspicion was aspergillosis. Upon reading the CW smear made from the specimen, septate hyphae earlier seen in the Gram smear did not show a dichotomous pattern of branching; instead these hyphal fragments appeared as thin and long and stick-like (*B: Image set 9.9*). Wet prep of the specimen examined under the bright field displayed thin, pale, dark hyphae, narrow but with parallel width, and the suspicion of a dark fungus involved was made (*C: Image set 9.9*). Fresh smears were made and sent to histology for GMS and melanin stain. GMS clearly showed thin fungal hyphae with septation against excellent background contrast. No branching pattern was observed. FM stain showed excellent melanin pigment in the fungal cell wall (*D: Image set 9.9*). About 7 to 10 days later, dark fungus started to appear on a mycology culture media (*E: Image set 9.9*). A quick scotch prep using LPAB was prepared, and structures producing conidia overlapping with *Fonseceae* were observed (*F: Image set 9.9*). The isolate was sent to the reference laboratory, where it was identified as *Ramichloridium mackenzeie*.

In the case above, structures seen were found to be unique in their own way. Very long, thin hyphae were seen and rare or no branching indicate fungi other than *Aspergillus* (personal experience). Rare branching forms crossing the boundary of the microscopic field must have confused the smear reader to name the structures as pseudohyphae, since pseudohyphae usually produce shorter filamentous forms attached to the adjoining cells in narrow base budding form. Although no pigmentation was found in the cell wall upon reviewing the Gram smear, pale-staining hyphae were observed under the bright field using wet prep. The structures seen were suspected of having dark fungus, and the specimen was stained by Fontana Masson and was positive for melanin pigment in the fungal cell wall. The fungus was identified as *Ramichloridium mackenziei* by experts in the reference lab.

Image set 9.9

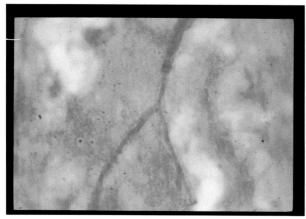

9.9-A Brain Abscess: Gram x1000 septate hyphae originally reported as pseudohyphae

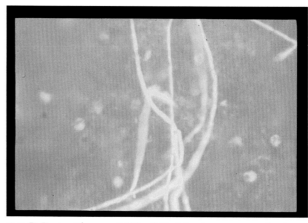

9.9-B Brain Abscess: CW x400 thin, long septate hyphae, rare branching

9.9-C Brain Abscess: WP x400 pale thin fungal elements clearly seen in septate morphology and appear pale

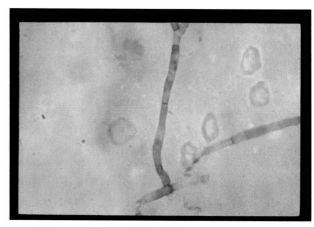

9.9-D Brain Abscess: FM x1000 brown septate hyphae indicating pigment of dematiaceous fungus

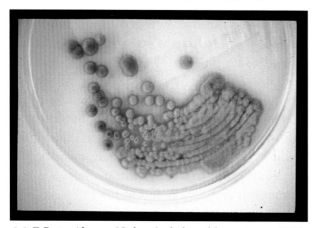

9.9-E Brain Abscess (Culture): dark mold growing on IMA

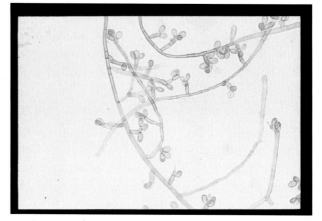

9.9-F Brain Abscess: LPAB x400 ovoid conidia resembling *Fonseceae*; identified as *Ramichloridium mackenzeie* by the reference lab

9.10 Case 41 – *Graphium basitruncatum*[22]

A 70-year-old leukemic patient was admitted to the hospital for induction chemotherapy. He also received itraconazole, an antifungal therapy, as prophylaxis. He failed to acquire adequate neutrophil recovery and was discharged. He was re-hospitalized after three weeks with febrile neutropenia. He had several non-tender, erythematous skin nodules along his extremities. He was treated with antibacterial agents. Two consecutive blood cultures were taken and processed in a routine microbiology lab that showed fungal elements that were reported out as yeast and pseudohyphae (*A: Image set 9.10*). The blood-culture bottle was subcultured on blood agar (BA) and was referred to the mycology section within the microbiology department for yeast identification. A subcultured blood agar plate was incubated at 30⁰C. No yeast was isolated from the subculture, but a dark fungus began to appear on the plate after few days (*B: Image set 9.10*). The dark fungus isolated was not identified by conventional testing using LPAB. Microscopic structures were seen under the microscope, but conidia produced by this dark fungus resembling several other fungal species and not fitting with any regularly known dematiaceous fungal species (*C: Image set 9.10*). The isolate was referred to the local reference lab, where it remained unidentified and was forwarded to another reference lab that identified the fungus as *Graphium basitruncatum*.[22]

The patient was started on voriconazole and caspofungin. Multiple blood cultures became positive, indicating that this dark fungus had spread through the blood stream to the skin, causing necrotic fungal nodules in this patient whose immunocompetency status had been compromised. The patient started to improve with liposomal amphotericin B, which was only temporary. The patient died from relapse infection.

The question comes to mind about the interpretation of the fungal elements seen and reported as pseudohyphae from the initial Gram-stained smear. Any fungal elements (true or pseudohyphae) growing in the blood culture are important to report as quickly as possible; however, the accuracy of the fungal elements seen must be carefully interpreted in order to relay meaningful information to the clinician, who would take immediate action to put the patient on appropriate antifungal therapy based on the kind of information the clinician received from the diagnostic laboratory personnel. We reviewed the original Gram-stained smears made from all positive blood-culture bottles and noticed that the elongated pinkish cells seen were not pseudohyphae. A considerably moderate amount of the fungal elements were present in the Gram-stained smears. Therefore, budding forms of the yeast must be searched to link pseudohyphae in the yeast entity, which was not in sight in any of the Gram smears made from positive blood-culture specimens. A closer look also displayed a "gap" between the two fungal filaments, without the "pinched-off" constriction that is often seen in pseudohyphae. The space that is seen in between the filaments indicates a septum between the two cells of a true hyphae. The Gram smear results were amended as "septate hyphae" seen.

Image set 9.10

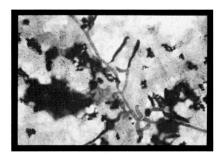 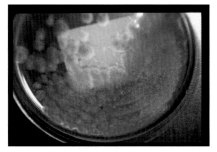 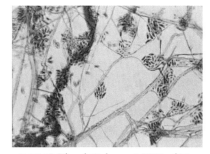

| 9.10-A Blood: Gram x1000 septate hyphae initially reported as yeast | 9.10-B Blood (Culture): growing dark fungus on IMA | 9.10-C Blood Culture: LPAF x400 identified as *Graphium basitruncatum* by the reference lab |

CHAPTER 10

10.1 ZYGOMYCETES[8, 9, 20, 23, 29, 33, 42]

Aseptate hyphae are 3–25 μm (three to five times the width of the septate hyphae) and are non-parallel in width. Septate hyphae more or less remain parallel, and aseptate hyphae (irregular width) are too wide in some areas and become too narrow at the end. The term commonly used for the clinical condition involving Zygomycetes is mucormycosis (zygomycosis), in which aseptate hyphae are seen in direct smears. Like other fungi, Zygomycetes are distributed in the environment and often isolated from clinical specimens as environmental contaminants. However, among immunocompromised hosts, Zygomycetes could have fatal consequences if left unattended.

10.2 CASE 42 – *RHIZOPUS*

A 43-year-old male lung-transplant patient had an abscess formation in the thigh near his groin area. Tissue from the thigh was received in microbiology for C&S and fungal investigation. Gram stain results were reported out as "no pus cells and no bacteria seen." Fungal smear stained by CW showed many aseptate hyphae suggesting Zygomycetes (*A: Image set 10.2*). The Gram smear was reviewed and found loaded with aseptate hyphae (*B: Image set 10.2*). However, the aseptate hyphae seen were so bizarre that it required expertise to correctly categorize them.

Unless the smear reader is aware of the nature of aseptate hyphae, it could be easily missed or remain undetected in Gram-stained smears until hunting them down with a passionate approach to detecting them. Undetected Zygomycetes in direct smear specimens may have fatal consequences in immunocompromised patients. Since this fungus has the propensity to invade the arterial blood system, it can clog up the arteries, causing death within a few days if it remains untreated. The culture grew as a fluffy, woolly, white, and grayish mold that turned darker, having a "salt and pepper" appearance with age (*C: Image set 10.2*). *Rhizopus* species was identified by microscopic morphology seen in LPAB preparation (*D: Image set 10.2*).

Zygomycetes may also be found in culture media as contaminants. Since the technologists are not fully aware of the clinical setting or other data to determine the clinical significance of Zygomycetes, it is safe to notify the ward or the physician looking after the patient having zygomycosis.

Aseptate hyphae present in the Gram smear are usually recognizable. However, careful observation is needed for the clue to find the fungal elements that remain unregistered in the mind but are visible under the microscopic field. Therefore, when reading direct Gram smears always observe structures that comprise the morphology. Then try to classify those structures into categories until you are able to differentiate in order to identify microorganisms accurately. The Gram in this specimen displays structures that are not clearly fungal elements. However, if you would walk along the tracks, you would be surprised to see that an image of fungal hyphae could at least be suspected. At this point you would attempt to make another smear, not a fresh Gram but a simple wet prep using KOH or CW/KOH. The CW/KOH prep would be more acceptable, since the same field of the specimen smear can be examined under the microscope as fluorescent stain and unstained KOH. Observe the Gram image carefully and form an outline along distorted fungal elements, looking like debris and appearing as twisted and intertwined hyphal-like fragments trying to confuse the smear reader. This is nothing but excellent aseptate hyphal fragments that need to be brought to the clinician's attention immediately.

Image set 10.2

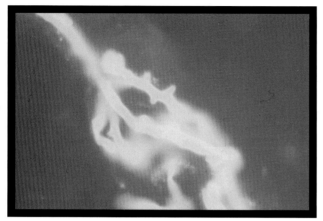

10.2-A Groin Tissue: CW x400 broad ribbon type aseptate hyphae, beginning to produce right angle branching

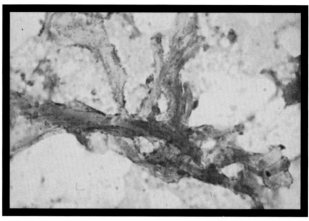

10.2-B Groin Tissue: Gram x1000 broad ribbon type aseptate hyphae, some aseptate hyphae display narrow width

10.2-C Groin Tissue (Culture): woolly, fluffy and greyish mold on IMA

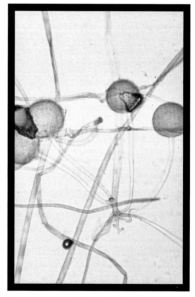

10.2-D Groin Tissue: LPAB x100 nodal rhizoids, no apophysis, round sporangia identified as *Rhizopus* species

10.3 CASE 43 – *ABSIDIA*

Patient data is unavailable except for the nature of the specimen and the fungal species isolated. A wound specimen from a patient was received in microbiology for C&S and fungus. "Pus cells seen and no bacteria seen" was reported to the clinician. In the mycology section excellent fungal elements of aseptate morphology were observed (*A: Image set 10.3*). The Gram smear was reviewed, and aseptate hyphae were observed (*B:*

Image set 10.3). A rapid-growing silver grayish mold was recovered from the corresponding culture (*C: Image set 10.3*) and was identified as *Absidia* species using LPAB (*D: Image set 10.3*) and (*E: Image set 10.3*)

Upon finding aseptate hyphae in the fungal smear, the Gram smear was reviewed. Excellent structures of aseptate hyphae were observed that were not picked up by the smear reader. This smear demonstrated unusual clarity with excellent contrast. Structures depicted no doubt about fungal elements, although they were huge in the image (under the microscopic field); however, the unique morphology suggested that the smear reader should bring this image to his/her supervisor's attention. When an object presents a definitive morphology, it must be followed through until identified. Unreported fungal elements of this nature could produce undesirable outcomes.

Attention must also be paid to the CW image. It has all that is required for identifying aseptate hyphae, such as non-septate hyphal fragments, wider, like ribbon in some parts, twisted and narrower at one end. Many texts describe the pattern of branching at right angles for aseptate hyphae. This is partially true and must not be hooked up to look for right-angle branching hyphae in aseptate hyphae. Many times right angles are present, but they are not clearly observed in the specimen smears most of the time (personal experience), because these fragile hyphae without septation often get tangled or clogged up and are unable to demonstrate their unique morphology as suggested.

Image set 10.3

10.3-A Wound Swab: CW x400 aseptate hyphae

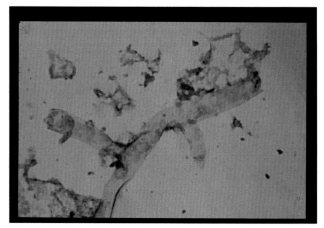

10.3-B Wound Swab: Gram x1000 aseptate hyphae displaying right angle branching.

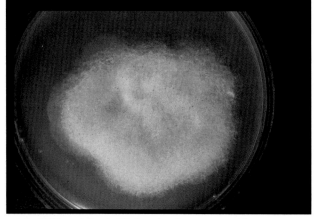

10.3-C Wound Swab (Culture): silver grey mold on IMA

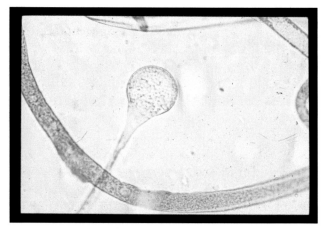

10.3-D Wound Swab: LPAB x100 pyriform sporangia with prominent funnel-shaped apophysis identified as *Absidia corymbifera*

10.4 CASE 44 – MUCOR

Patient data is unavailable at the moment except for the nature of the specimen and the fungal species isolated. An antrum biopsy specimen was processed for C&S and fungal investigation in microbiology. The Gram smear results reported out as "pus cells and commensal flora seen." A fungal smear (CW) done in mycology showed excellent structures that also generated some ideas (*A: image set 10.4*). C&S and the mycology culture rapidly grew fluffy mold that quickly turned grey (*C: Image set 10.4*). Lactophenol prep using LPAB showed round sporangia filled with sporangiospores, and no rhizoids were seen (not shown in the image). The structure shown in the microscopic image is showing excellent knobby columella of *Mucor* (*D: Image set 10.4*). Examine the CW image and observe the branching pattern. The structures seen in this image appear to be dichotomous; i.e., acute-angle branching. However, the branching seen in the image are right-angle branching and have been drawn inward due to mechanical manipulation. A closer look describes two important points: 1) the structures at this magnification are much wider than dichotomous branching septate hyphae; and 2) the curled-up hyphae have no septum. The branching pattern is true right-angle branching but does not demonstrate the textbook description.

Upon review, the Gram-stained image is also interesting and different from other aseptate hyphae shown in the above cases (*B: Image set 10.4*). The image appears as pink "brushed stripes" on the wall without proper definition. Aseptate hyphae often display an uneven staining pattern totally undifferentiable from staining reagents or specimen debris. However, moving eyes along the tracks begins to create an image that helps the smear reader to do some "fill-in the blank" strategy in order to identify structures as aseptate.

Image set 10.4

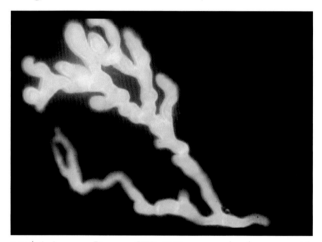

10.4-A Antrum Biopsy: CW x250 aseptate hyphae, too wide at some points and narrow in other areas

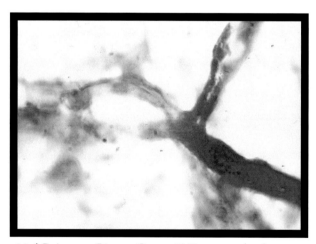

10.4-B Antrum Biopsy: Gram x1000 aseptate hyphae, too wide at some points and narrow in other areas

10.4-C Antrum Biopsy (Culture): rapidly growing woolly and greyish mold on SAB

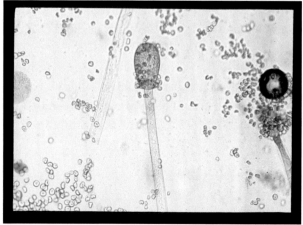

10.4-D Antrum Biopsy: LPAB x400 knobby columella, no rhizoids, round sporangiophores (not shown in the image) identified as *Mucor* species

Chapter 11

11.1 DERMATOPHYTES[8, 9, 20, 23, 29, 33, 42]

Dermatophytes are fungi causing superficial infections of the hair, nail, and skin. There are three genera in this group; *Microsporum, Trichophyton,* and *Epidermophyton. Microsporum* causes infections of the hair and skin but not nail. *Trichophyton* caused infections of the hair, nail, and skin. *Epidermophyton* causes infections of the nail and skin but not hair. Specimens for the recovery of dermatophytes are collected carefully in folded black paper or in a sterile container. Infected hair, nail clippings, and skin scrapings are often sent to microbiology for isolating dermatophytic agents causing superficial mycosis. Certain fungi other than dermatophytes, such as *Scytalidium, Onychocola, Malassezia,* and others are also known to cause superficial mycosis.

11.2 CASE 45 – *TRICHOPHYTON TONSURANS* IN CORNEA

Cornea scraping from a female patient suffering from pain and redness of the eye was received from the eye clinic for C&S and fungal investigation. "Pus cells seen and no bacteria seen" was reported from Gram. CW done in mycology showed short, small, and elongated fungal elements not falling into any specific category (*A: Image set 11.2*). The Gram smear was reviewed, and fungal elements were observed (*B: Image set 11.2*). C&S reported "no growth after a week," and culture media was discarded as per departmental policy. A mycology culture media was in progress that began to show a low, pale, white granular mold (*C: Image set 11.2*). The mold turned tan, beige, and brown at maturity. A wet prep using LPAF was made, and *Trichophyton tonsurans* was identified (*D: Image set 11.2*).

This case is extremely interesting for isolating a fungus that is not commonly seen in cornea specimens. Dermatophytes are superficial mycotic agents and do not usually infect the cornea. The eye is not protected in an enclosed cavity; however, its site and the nature of tissue material do not contain the keratinized dead tissue or decaying organic matter that dermatophytic agents love to feed on. Fungal elements seen did not give any indication about what the corresponding culture isolate would be. The Gram and CW smears show single fungal elements, some oval, other elongated or club-shaped. The morphology was a bit closer to the yeast or yeast-like fungi. However, there was no evidence to place these fungal elements in any particular entity. The patient was treated successfully.

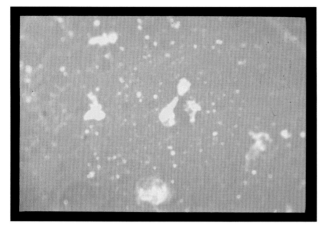

11.2-A Cornea: CW x400 fungal elements seen

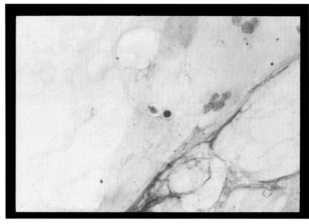

11.2-B Cornea: Gram x1000 fungal elements and two-yeast-like cells seen

11.2-C Cornea (Culture): sued-like slow growing white mold (turning beige) on IMA

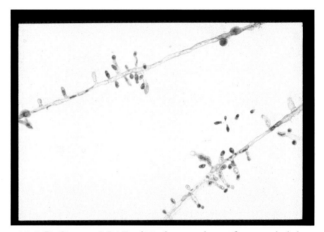

11.2-D Cornea: LPAF x400 elongated, pyriform and club-shaped microconidia of *Trichophyton tonsurans*

11.3 CASE 46 – *TRICHOPHYTON TONSURANS* IN EAR

Otomycosis is usually caused by *Aspergillus niger*. However, any opportunistic fungus can cause infection in the ear. Certain fungi are classified based on the specific nature of the fungal infections they produce; for example, superficial mycotic agents are capable of causing fungal infections on surfaces of the human body and do not dig deep inside the human tissue. Dermatophytes are a group of fungi that cause infections of the skin, hair, and nail.

An ear swab was received from the ENT clinic requesting C&S and fungal culture. The Gram smear was reported as "pus cells seen no bacteria seen." A CW smear in the mycology section showed fungal elements of septate morphology. The Gram smear was reviewed and showed fungal elements consistent with true hyphae (*Image set 11.3*). A bacterial culture report was sent as "normal flora after 48 hours." The mycology culture began to show fungus after one week. The culture matured in 10 days by producing low mycelium, pale, beige and granular morphology. Wet prep using LPAB was made and examined under the microscope. The structures seen under the microscope identified the fungus as *Trichophyton tonsurans*. The culture turned brown on reverse and deep tan and granular on the surface with age.

Many laboratories might consider the identification of the fungus isolated incorrect or the specimen may have been mislabeled since dermatophytes only cause infections on the superficial sites and not inside the human body. The reason for dermatophytic agent to infect the ear canal is that otitis externa is a superficial site. The fungus was not recovered from deeper sites. Therefore it is possible that superficial mycotic agents such as dermatophytes may be isolated from the surface of the ear canal. However, the recovery of dermatophytes from such sites is uncommon.

Image set 11.3

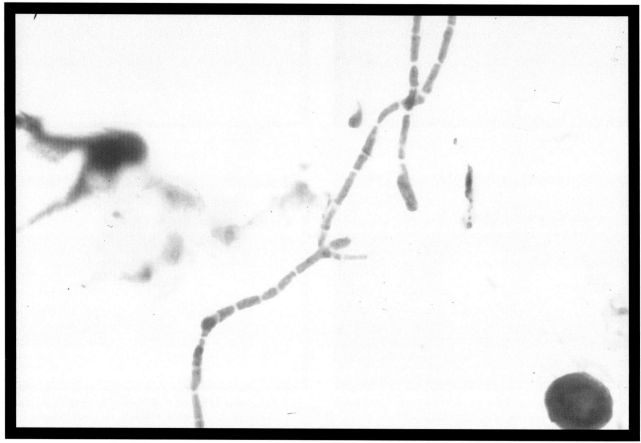

11.3 Ear: Gram x1000 septate hyphae breaking up into arthroconidia, *Trichophyton tonsurans* was recovered from culture.

11.4 CASE 47 – *TRICHOPHYTON TONSURANS* FROM SCALP

A swab specimen from a lesion on the scalp was collected from a patient and sent to microbiology for C&S and fungal culture. The Gram smear results from C&S were reported as "pus cells and yeasts seen" (*A: Image set 11.4*). A CW smear in mycology showed filamentous forms in septate hyphae morphology (*B: Image set 11.4*). The Gram smear was reviewed, and it was noticed that the smear reader must have been misled by the round structures sitting side by side mimicking budding yeast. The corresponding fungus culture grew mold in about seven days, identified as *Trichophyton tonsurans* using lactophenol aniline blue preparation.

This image demonstrates the mechanism of a Gram reaction, which shows purple color for most yeasts seen in direct Gram-stained smears. The structures seen in the image do not belong to a yeast entity. A closer look would show that the two round structures are not budding but are attached by the septum and not pinched-off structures of budding yeast. Surrounding the round structures, there is a pinkish background indicating cytoplasm of the fungal cell whose boundary (cell wall) is not defined or fully resolved. The attachment located centrally between the two cells is believed to be a septum whose ends are not stained purple by the CV reagent. A CW image of the same specimen is showing fungal hyphae in septate morphology. Culture did not grow any yeast in C&S or mycology culture media. The clinical data also suggests that the lesion on the scalp (superficial site) is most likely to occur due to a dermatophytic agent.

When reading direct smears searching for microorganisms, the smear reader must keep looking for clues based on the clinical data of the patient, the structures seen making a definitive morphology, and the details that the microscopic objects display in order to identify microorganisms correctly and assign them to the specific categories by interpreting fungal elements precisely so that the clinician receives meaningful information for the proper management of the patient.

Image set 11.4

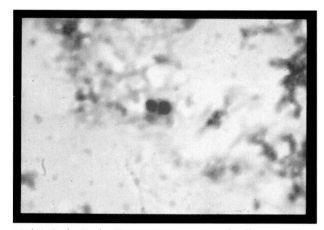

11.4-A Scalp Swab: Gram x1000 two round cells reported as yeast from Gram stain

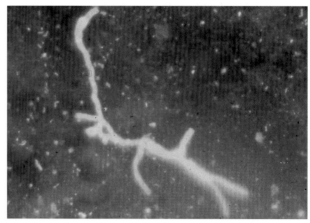

11.4-B Scalp Swab: CW x400 fungal elements appear to be septate. Culture recovered from this specimen is *Trichophyton tonsurans*

Chapter 12

12.1 Yeast/*Candida*

Yeast/*Candida* are unicellular organisms, produced asexually by budding, fission or a combination of both. Yeasts producing elongated cells without separating from the mother cell and appear in chains are known as pseudohyphae[9]. The term *Candida* applies to yeasts producing pseudohyphae. Certain yeasts are able to produce true hyphae while other yeasts produce different structures.

Yeasts are ubiquitous in the environment and are commonly isolated from clinical laboratories.[8] They are also normal flora of the body; therefore it is sometimes difficult to determine the clinical significance, especially when yeasts are isolated from surfaces. It is therefore very important to demonstrate yeast in direct microscopy in order to determine clinical relevancy. The morphological structures of certain yeasts are species-specific. The presence of yeast in the clinical specimen observed under the microscope guides the reader to expect to recover correlating yeast from the corresponding culture medium.

A direct smear stained by Gram method detects yeast easily. Structures seen under the microscope in direct smears resembling yeast without being isolated from the culture media may alert the smear reader to extend the incubation of planted media. And yet in another scenario a yeast-looking organism isolated from culture may be easily ruled out after observing morphological structures in the direct microscopy.

Yeasts and *Candida* are identified by the shape and the size of the specific structures they produce. Yeasts are oval to round cells (2–20 μm)[8] some of which may produce filamentous forms such as pseudohyphae.

12.2.1 Case 48 – *Candida albicans*

Different clinical specimens (A: Abscess; B: Blood; C: Mandible lymph node; D: Eye fluid) were collected over a period of six months from patients suspected of having fungal infections. The specimens were processed for culture, and direct Gram smears were examined. The Gram smear images posed some problems for the smear reader. Two of the four images were reported as "no bacteria or fungal elements seen" (*B & D: Image set 12.2.1*), and the other two images were assessed by the expert technologist who decided to report pseudohyphae from abscess (*A: Image set 12.2.1*) and mandible lymph node (*C: Image set 12.2.1*). Images A & C are important in terms of Gram reaction. Image A and image C both appear to be Gram negative. However, a closer look at image A demonstrates that the pink stain (safranin) is sticking to the outer boundary of the cell wall. It appears that all basic dyes used in the Gram stain are prevented from entering the pseudohyphae, as a result helping easy detection due to the contrast produced by safranin. This image is not stained with Gram reagent, proving a hypothesis that we have earlier stated, that the cell wall of fungi may not allow the primary reagent to get in due to narrow pores unable to let a primary reagent enter the cell. In image C, the Gram reaction displays a pinkish tinge in the interior of the yeast cell. It is still not clear if the cell allowed entry of CV that got washed away during decolorizer use, but it at least demonstrates that the decolorizer must have caused some damage to the cell wall, allowing a counterstain such as safranin to get in. Therefore, the interior of the cell shows a pinkish tinge. In both images, the contrast and the staining pattern created a boundary around the objects that appeared as pseudohyphae. The culture belonging to image A grew yeast that was identified as *Candida albicans*. However, no growth was obtained from the specimen (image C) after four weeks. DNA studies identified the yeast as *Candida albicans* done at the reference laboratory.

Image set 12.2.1

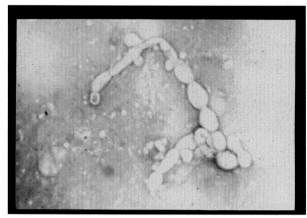

12.2.1-A Abscess: Gram x1000 unstained pseudohyphae

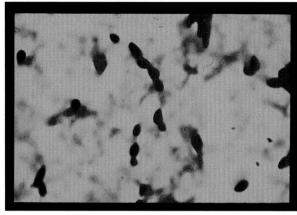

12.2.1-B Blood: Gram x1000 yeast cells staining purple

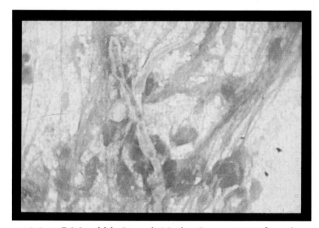

12.2.1-C Mandible Lymph Node: Gram x1000 fungal elements showing pinkish interior of the pseudohyphae not clearly standing out due to poor contrast

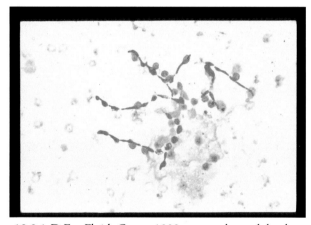

12.2.1-D Eye Fluid: Gram x1000 yeast and pseudohyphae are purple and clearly visible

12.2.2 CASE 49 – *CANDIDA ALBICANS*

Another set of four clinical specimens (A: Eye fluid; B: Cornea; C: Vitreous; D: Pleural fluid) processed in microbiology from four different patients also grew *Candida albicans*. The set of images in this case was not straightforward. That's where imagination kicks in to play a role. Three of the four specimen images belong to the similar site involving the eye. Self questioning and answering leads to a way to do a differential diagnosis of the structures observed under the microscope. Precise interpretation was prepared for the clinicians in a timely fashion so that appropriate therapy may be initiated for the proper management of the patient.

Gram smears from the above set are interesting. Each one has its own peculiarity in terms of confusion but provides a clue for solution as well. In the first image (*A: Image set 12.2.2*) long, thin-walled, stick-like structures with variable staining patterns are observed. It is easy to rule out these structures as true hyphae since there is no septum seen but constrictions at segments are observed. No single or budding yeast in sight, that's a bit odd for yeast/*Candida* category. However, the presence of pseudohyphae structures in the Gram smear provide sufficient evidence to include them in the yeast or yeast-like fungus. The corresponding culture grew *Candida albicans*. The image (*B: Image set 12.2.2*) showing unclear structures that are difficult to categorized. The nature of the specimen is such that anything seen in the direct smear must be followed through until identified since unreported and unidentified microorganisms could remain undiagnosed, compromising patient care. Structures seen in this image are unstained and in filamentous forms. The width of the objects is greater than 1 μm; non-branching and elongated cells stacked side by side producing doubtful morphology. The decision was made to term these structures "fungal elements seen." The corresponding culture grew *Candida albicans*. The image (*C: Image set 12.2.2*) from the vitreous fluid falls into the pseudohyphae morphology. The yellowish-tinged specimen material interferes with contrast in the Gram smear, compromising image quality and clarity. The last image in the set (*D: Image set 12.2.2*) is straightforward, demonstrating round oval cells and the budding form. The cells do not show strong Gram reaction and also appear more like conidia than yeast. However, clear budding form was observed. The corresponding culture grew *Candida albicans*.

Image set 12.2.2

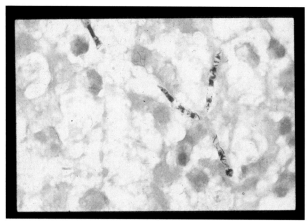

12.2.2-A Eye Fluid: Gram x1000 Eye pseudohyphae stained & unstained, culture recovered *C. albicans*

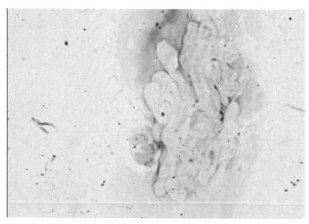

12.2.2-B Cornea: Gram x1000 unstained stacks of fungal elements in pseudohyphae morphology, culture recovered *C. albicans*

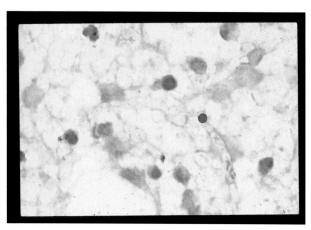

12.2.2-C Vitreous Fluid: Gram x1000 unstained pseudohyphae some show yellowish tinge, culture recovered *C. albicans*

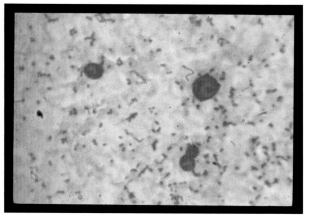

12.2.2-D Pleural Fluid: Gram x1000 two single yeast-like cells and a budding form, culture recovered *C. albicans*

12.2.3 Case 50 – *Candida (Torulopsis) glabrata*

Two clinical specimens (A: Kidney aspirate; B: Peritoneum biopsy) were received from two different patients in separate events. Direct Gram smears were performed and "NBS" were reported. The fungal smear (CW) showed small oval budding yeasts in both smears. The corresponding culture grew yeast, identified as *Candida (Torulopsis) glabrata*. The Gram smears were retrieved and examined under the microscope. The structures seen (*A: Image set 12.2.3*) are not clear. As a result, fungal elements were missed by the smear reader. Careful observation allowed us to spot unstained objects located centrally. The clarity of these objects was compromised by the lack of contrast. As a result the smear appeared negative for fungal elements. The cells sitting in the center of the microscopic field appear oval/round in morphology. Other characteristics are missing due to resolution difficulties. However, the shape and size of the structures categorize these objects as fungal cells in yeast morphology, which was later confirmed upon isolating *Candida glabrata* from the corresponding culture. The Gram smear from peritoneum (*B: Image set 12.2.3*) was not a problem to locate single cells smaller in size and oval in shape. Although no budding forms were seen, structures seen were recognized as yeast. It must always be remembered that structures resembling yeast must be confirmed by budding form before reporting. Many times yeasts appear single. Therefore, keep searching other microscopic fields to demonstrate budding at some point even if only a few budding forms are seen. In cases where the true budding forms are not seen, do not call the structures yeast without a statement about what these structures could be. Culture confirmation is essential. *Candida glabrata* usually produces small sized oval yeast cells (3 to 5 μm) of the same morphology as *Histoplasma*. *Histoplasma* is usually found intracellularly and stains poorly with Gram's method. Similarly *Candida glabrata* usually stains purple (Gram positive) and may rarely be seen intracellularly.

C. glabrata is the most common non-*C. albicans* in North America (2001–2004).

Candida guilliermondii is resistant to resistant to amphotericin B, fluconazole.

Candida rugosa, showing increasing fluconazole resistance 30%→ 60%; voriconazole 3.1% to 38%.[15,37,38] *Candida krusei* is intrinsically resistance to fluconazole; *Candida lusitaniae* is usually resistant to Amphotericin B. Fluconazole is useful for *C. albicans* and *Cryptococcus neoformans*.[7]

Image set 12.2.3

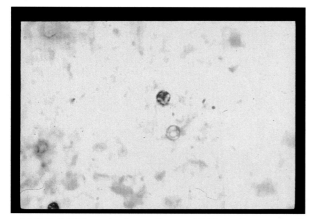

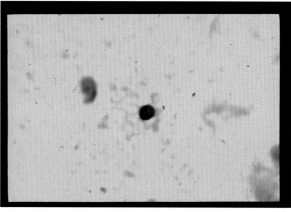

12.2.3-A Kidney Aspirate: Gram x1000 single round cells, one of the two single cells appears to produce a bud, culture recovered *Candida glabrata*

12.2.3-B Peritoneum Biopsy: Gram x1000 single, round cell, difficult to name it as a yeast or conidium, culture recovered *Candida glabrata*

12.3 CASE 51 – *CRYPTOCOCCUS NEOFORMANS*

A 54-year-old male renal transplant patient had a short history of headache, productive cough but no fever. CT scan of chest showed right pleural density. Lung biopsy specimen was sent to histopathology, where a diagnosis of *Histoplasma* was made. However, the clinician wanted clarification, and the GMS smear from pathology was forwarded to mycology for review. Upon examining GMS smear under the microscope it was found to be loaded with clusters of round cells that did not fit into the morphology of *Histoplasma*. A lung biopsy specimen was retrieved and processed for CW/KOH prep and examined under the UV as well as bright-field microscope. Visual clarity was greater under the fluorescent light, and the majority of cells observed were round. Upon viewing the smear under the bright field, round cells appeared having slightly darker cell walls consistent with *Cryptococcus*. Since the histopathology diagnosis was *Histoplasma*, we suggested to the clinician that if they ever decide to draw CSF specimen from the patient, it must be checked for the cryptococcal antigen. Immediate action was taken, the CSF specimen was drawn and sent to the microbiology lab for the cryptococcal antigen that turned out to be positive (1:64). A subsequent culture grew *Cryptococcus neoformans*, and the antifungal agent was changed to liposomal amphotericin B.

The clinician asked mycology's opinion on the objects seen in the GMS smear and found interesting organisms. The Gram smear was reported as "pus cells seen and no bacteria seen." The fungal smear (CW) showed yeast cells that were essentially round and not fitting in *Histoplasma* entity (*A: Image set 12.3*). The Gram smear was reviewed and noticed single, thick-cell-walled and irregularly round cells (*B: Image set 12.3*). The structures seen in the Gram smear are unusual and may have confused the smear reader with tissue cells or artifact. Wet prep using KOH was done on the culture and yeast cells found that were consistently round and had darker cell wall (*C: Image 12.3*). India ink done on culture shows excellent capsulation of *Cryptococcus neoformans* (*D: Image 12.3*). Cornmeal agar morphology displayed excellent structures of round yeast cells, some having space in between (due to capsulation) and having the darker cell wall that is the most important characteristic often seen in *Cryptococcus neoformans* (*E: Image set 12.3*). The specimen was also stained by mucicarmine to demonstrate capsule staining. The capsule and the yeast cell of *Cryptococcus neoformans* both stain pink (red) by mucicarmine (*F: Image set 12.3*). Care must be taken to correctly identify *Cryptococcus neoformans*, separating it from *Blastomyces*, which is also stained pink by mucicarmine. Culture grew *Cryptococcus neoformans*. No *Histoplasma* was isolated.

Image set 12.3

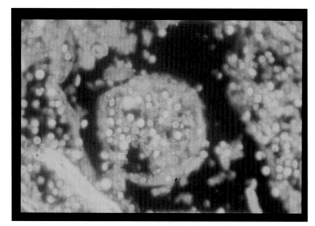

12.3-A Lung Biopsy: CW x400 round cells, mostly single suspected of *Cryptococcus* (pathology reported *Histoplasma*)

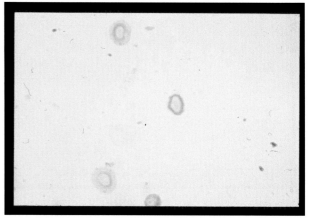

12.3-B Lung Biopsy: Gram x1000 near spherical thick-walled and appear to be masked by safranin around the outside the cell wall (encapsulated *Cryptococcus*, pathology reported *Histoplasma*)

12.3-C Lung Biopsy: WP x400 round and dark cell-walled Cryptococcus (pathology reported *Histoplasma*)

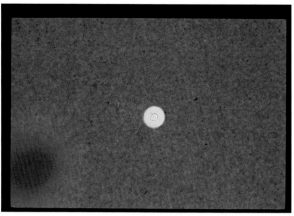

12.3-D Lung Biopsy: India ink x400 demonstrating the presence of capsule surrounding *Cryptococcus*

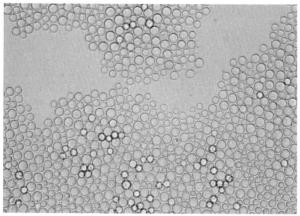

12.3-E Lung Biopsy: Cornmeal Agar x400 dark cell-walled round cells of *Cryptococcus* sitting at space indicating the presence of capsule are clearly observed

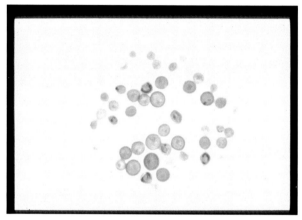

12.3-F Lung Biopsy: Mucicarmine x1000 thick cell-walled (due to accumulated gelatinous material) and round cells staining pink (body and the capsule of *Cryptococcus neoformans* stain pink by Mucicarmine). Care must be taken not to confuse with single cell *Blastomyces* that also stains pink with Mucicarmine.

12.4 CASE 52 – *SPOROBOLOMYCES*

Sporobolomyces salmonicolor is yeast that ejects blastospore forcibly, which lands at a distance, giving the appearance of scattered colonies looking like contamination. *Sporobolomyces* is usually nonpathogenic. It may cause infections among immunocompromised hosts. We have received a CSF specimen in microbiology from a debilitated person whose immune status was partially suppressed due to medication. CSF grew coral-colored yeast (confirmed by wet prep) on inhibitory mold agar (IMA), assuming it to be *Rhodotorula* by morphology (*A: Image set 12.4*). A Gram stain was performed on the colony and showed elongated yeast cells leaving behind a thin filament attached to one end (*B: Image set 12.4*), indicating a remnant left behind after the blastospore was ejected. Further tests to identify yeast including cornmeal agar (with Tween-80)[21] and PDA (*C & D: Image set 12.4*) were set up. At this point a double plate (one of which is streaked) facing the surface of each other were set up (*E: Image set 12.4*) to prove the ballistic blastospores settling opposite on the surface facing the streaked plate. Upon maturity, the two plates were unsealed and examined for satellite colonies (*F: Image 12.4*). Ejected blastospores were shot and settled on the surface of the opposite un-streaked plate. The letters "SKM" are showing similar orientation of the growth on the opposite plate confirming satellite colonies caused by the ballistic blastospores.

Image set 12.4

12.4-A CSF (Culture): red pigment producing yeast on IMA was suspected of *Rhodotorula*

12.4-B CSF: Gram stained smear was made form culture and examined at x1000 magnification that showed elongated yeast cells and a filamentous remnant left behind at one end giving the appearance of "rat tail" suspecting *Sporobolomyces*

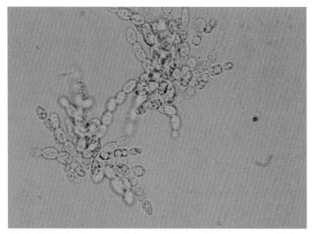

12.4-C CSF: Cornmeal Agar x400 microscopic morphology of the yeast-like fungus showing large cells in chains not fitting into any known Yeast/ *Candida* morphology.

12.4-D CSF (Culture): the organism produced intense coral red pigment on PDA

12.4-E Double Plate System: demonstrating ballistic conidiation

12.4-F CSF (Culture): the mirror image (satelliting colonies) produced by ejecting ballistic spores of *Sporobolomyces*.

12.5 CASE 53 – *GEOTRICHUM CLAVATUM* (?*BLASTOSCHIZOMYCES*)

A 63-year-old patient, severely immunocompromised, was admitted to the hospital with high fever and respiratory problems. A blood specimen was drawn in a bottle containing commercial media for the recovery of microorganisms and sent to the microbiology department for C&S. The blood culture-machine registered the blood-culture bottle as positive within three days. A direct Gram smear was made from the positive blood-culture bottle and read. Yeast and pseudohyphae (*A: Image 12.5*) were observed in the Gram smear and reported on line. Yeast isolated turned mycelial around the periphery of the moist colonies on blood agar (*B: Image 12.5*) and remained moist, pale on the chocolate agar, but growth on SAB was dry. Yeast identification tests were set up (cornmeal agar commercial sugar assimilation test and slide culture). A sugar assimilation test gave two choices, with the highest being *Trichosporon capitatum* (86%) and *Candida krusei* (6.9%). *Candida krusei* was rejected based on the microscopic morphology produced on the cornmeal agar. Cornmeal agar morphology displayed excellent structures (*C & D: Image 12.5*) that resembled *Blastoschizomyces* (formerly called *Trichosporon capitatum*). Slide culture wet prep made in lactophenol showed elongated cells (*E & F: Image 12.5*) different than pseudohyphae and not matching regular arthroconidia usually seen in *Trichosporon* or *Geotrichum*. The isolate was sent to the reference lab for complete analysis, and the results came back from the reference lab as *Geotrichum clavatum*. We did not agree with the results; however, the *Geotrichum clavatum* result reported by the reference lab was posted on line. Sometime later, molecular techniques were applied to the specimen, which identified the yeast in question as *Trichosporon capitatum*, now known as *Blastoschizomyces capitatum*. The separation of this species from the genus *Trichosporon* is based on the structures it produces, which are known as "annellides."[17, 24] In routine practice, it is very hard to assess annellidic structures produced by *Blastoschizomyces*. However, routinely examining cornmeal morphology of all yeast is essential since it confirms the yeast identification achieved by commercial products that cannot be used as a single test-based scenario. Unless there is some sort of back-up system, we would risk misidentifying a yeast only because other tests supporting identification are not immediately available to challenge the credibility of a test that is likely to fail. *Trichosporon capitatum* (*Blastoschizomyces*) is a known opportunistic pathogen causing systemic infections among immunocompromised hosts.

Image set 12.5

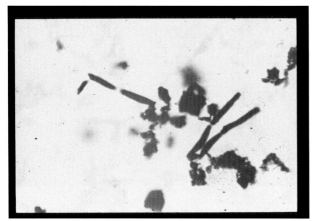

12.5-A Blood: Gram x1000 yeast/ pseudohyphae was reported initially. Upon review of the Gram smear; septate hyphae were seen.

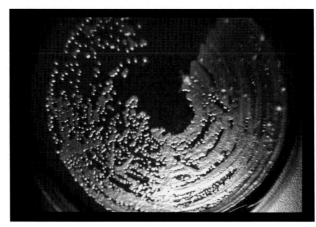

12.5-B Blood (Culture): yeast like fungus growing on chocolate agar

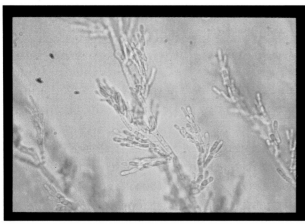

12.5-C Blood: Cornmeal Agar x400 microscopic morphology on the cornmeal agar resembled *Trichosporon* (annelloconidiation)

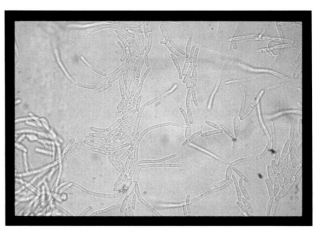

12.5-D Blood: Cornmeal Agar x400 second set of cornmeal agar showed structures resembling *Trichosporon/ Blastoschizomyces* (annellidic conidiation suspected)

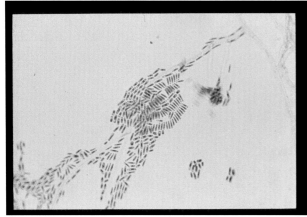

12.5-E Blood: LPAF x400 slide culture revealed contiguous arthroconidia resembling *Trichosporon/ Geotrichum*

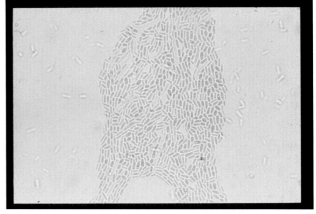

12.5-F Blood: LPAB x400 showing similar structures seen as mentioned above

CHAPTER 13

13.1 FILAMENTOUS ORGANISMS (AEROBIC ACTINOMYCETES AND MYCOBACTERIA) [8, 9, 23, 29, 33, 46]

The mycology lab within the microbiology department handles aerobic actinomycetes since their slow rate of growth and filamentous forms can be confused with mycelium. Mycobacteria are not particularly handled in mycology laboratory. However, when mycobacteria (mostly filamentous forms) are growing on mycological media, these organisms are identified to the genus level after running preliminary staining procedures to rule out *Nocardia* and other actinomycetes.

Nocardia is a bacteria and most probably will not appear on mycological media (unless resistant to antibiotics), since mycological media used to isolate fungi contain antibiotics. Most clinical specimens received for actinomycetes investigation are likely to contain bacterial contamination. Upon isolating bacteria-like growth on mycological media, *Nocardia* (other actinomycetes) and mycobacteria should be suspected.

Sometimes specimens received contain granules indicating mycetoma (bacterial-botryomycotic, actinomycotic, and eumycotic). All three types of mycetoma can be differentiated when stained by different staining procedures and examined under the microscope. Macroscopically, solid gritty sulfur granules are observed that are crushed between the two microscopic slides and stained appropriately in the microbiology laboratory in order to identify the organism causing mycetoma (*Image set 13.1*).

Image set 13.1

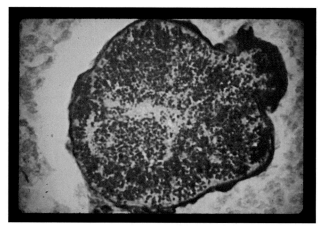 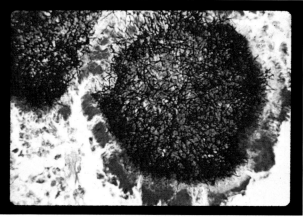

13.1-A Foot Abscess: Giemsa x400 Botryomycosis caused by bacteria (*Staph aureus*)

13.1-B Abdominal Wall: Gram x1000 sulphur granule, branching and clubbing formation (Actinomycotic)

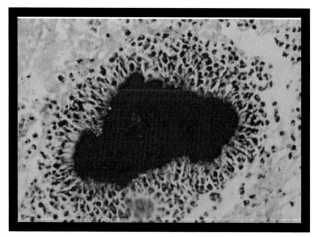

13.1-C Abscess: H&E x1000 sulphur granule, clubbing formation (Actinomycotic)

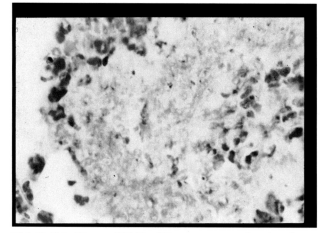

13.1-D Skin Biopsy: H&E x400 eumycetoma showing dense network of fungal hyphae in the periphery surrounding hyaline or dark hyphae seen in the central area.

13.2 CASE 54 – *NOCARDIA*

A sputum specimen from a female patient suffering from thymoma was received by microbiology for C&S and fungal investigation. The Gram smear results were reported as "pus cells seen and no bacteria seen." The fungal smear (CW) in mycology was also negative for fungal elements; however, thin branching bacilli with dull fluorescence were noticed. The Gram smear was reviewed and found to be Gram positive, with beaded and branching bacilli (*A: Image set 13.2*). Upon review, it was noticed that branching bacilli were present in the smear in moderate quantity; almost every other field contained these organisms, but the smear reader was unable to spot them. CW, on the other hand, does not detect bacteria that are much smaller (width < than 1 μm) than fungi. Bacteria do not stand out when stained by a fungal stain such as CW. The bacterial organisms in the fungal smear did not fluoresce but appeared in groups in thin thread-like morphology; the smear reader in mycology was able to suspect these structures as being meaningful, although not a fungal entity. A fresh smear was prepared and stained by modified kinyoun (MK) stain and partial acid fast-branching bacilli (*B: Image set 13.2*) were seen. The specimen was set up on special selective and differential medium (sodium pyruvate), and *Nocardia* was recovered within 10 days. Simple tests rules out non-pathogenic *Streptomyces* such as culture positive for partial acid fast by MK (*C: Image set 13.2*), ZN negative for acid-fastness (*D: Image set 13.2*), negative for casein hydrolysis (*E: Image set 13.2*), and fragmentation produced by slide culture (*F: Image set 13.2*). *Nocardia* is not a frequent occurrence in many laboratories around the globe. Therefore, certain routine microbiology laboratories run simple tests only to rule out nonpathogenic actinomycetes and to reach the genus level identification, pending speciation from the reference lab. *Nocardia asteroides* was suspected from our patient's sputum specimen and was confirmed by the reference lab. Had the mycology technologist examining the fungal smear (CW) not alerted the clinician about the possibility of *Nocardia* in the patient's specimen, the correct diagnosis would have not been established in a timely fashion, causing delay for appropriate therapy.

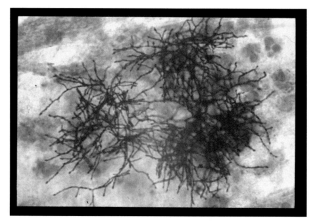

13.2-A Sputum: Gram x1000 Gram positive branching bacilli suggestive of *Nocardia*

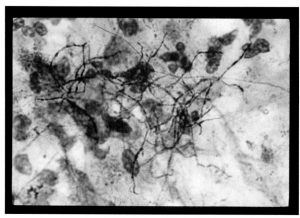

13.2-B Sputum: stained by Modified Kinyoun x1000 partial-acid-fast branching bacilli indicating *Nocardia*

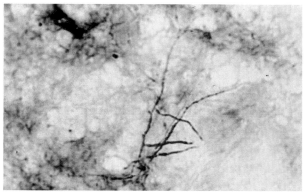

13.2-C Sputum culture: stained by Modified Kinyoun x1000 partial-acid-fast branching bacilli indicating *Nocardia*

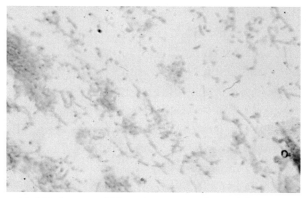

13.2-D Sputum: ZN x1000 negative for acid-fastness

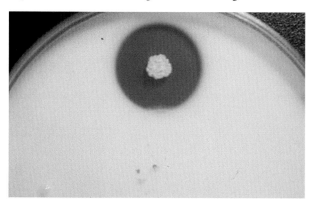

13.2-E Sputum: *Streptomyces* hydrolysing Casein (clear zone). No Casein hydrolysis observed by *Nocardia asteroids* (central area showing no clearing)

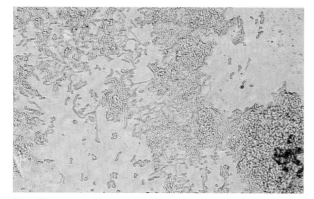

13.2-F Sputum: WP x1000 *Nocardia asteroids* fragmenting into bacillary form on PYRU medium

13.3 CASE 55 – *ACTINOMADURA MEXICANA*

A 34-year-old male patient had a lesion on his foot. A biopsy specimen was received for C&S, fungal culture, and mycobacteria, and *Nocardia/Actinomyces* investigation was specifically asked for. The Gram smear showed pus cells but no bacteria. The fungal smear (CW) was negative for fungal elements. A C&S culture showed no growth on plated and fluid media (aerobic and anaerobic) after seven days. The specimen was also inoculated on selective and differential medium sodium pyruvate (PYRU) [41] for the recovery of *Nocardia*. No growth was obtained from fungal culture media after four weeks. The PYRU medium started to show small pinpoint colonies after four weeks and slowly increased in size with extended incubation as raised, dry, pale, and irregularly folded in bacterial morphology (*E: Image set 13.3*). The color of the colonies also started to turn brown to light reddish and completely turned coral red upon maturity (*F: Image set 13.3*). After the recovery of glabrous bacterial colonies, Gram, MK, and ZN stain were done on the culture. The Gram smear reaction was pale and weak and showed thin branching bacilli (*A: Image set 13.3*). Another smear was prepared by crushing a small colony in between two slides and stained with Gram smear. This displayed better contrast and clearly demonstrated thin branching bacilli (*B: Image set 13.3*). The culture was stained by modified Kinyoun (MK) and Ziehl-Neelsen (ZN) stain, and both were negative for partial acid-fastness and acid-fastness respectively, ruling out mycobacteria and *Nocardia* (*C & D: Image set 13.3*). Based on the macroscopic and microscopic morphology and the coral red color, *Actinomadura* species was reported to the clinician and the isolate was sent to the reference lab for speciation. *Actinomadura mexicana* was identified by 16 S ribosomal sequencing at the reference lab. The original Gram smear was discarded after two weeks as per departmental policy and unavailable for review. *Actinomadura* grew very slowly after four weeks.

Mycetoma is a chronic, granulomatous disease of the skin and subcutaneous tissue and may involve muscle, bones, and neighboring organs. Typically affecting the lower extremities, it can also occur in almost any region of the body. Gill first described the disease in the Madura (Mathura) district of India in 1842 and termed the clinical condition Madura foot. Mycetoma is produced by direct local trauma to the skin with thorn, wood splinters, and other objects. Clinically, the disease begins as small, firm nodules and progresses to form extensive suppurative plaques. Eumycetomas tend to be more localized than actinomycetomas. The grains are surrounded by polymorphonuclear cells (PMNs), lymphocytes, plasma cells, and histiocytes. Mycetoma is more common in men than in women. The histopathologic picture reveals a granulomatous inflammation with abscess formation. A central zone exists where polymorphonuclear cells are abundant and granules or grains are found. This central zone is surrounded by lymphocytes, plasma cells, histiocytes, and fibroblasts. The actinomycetes are seen by staining with hematoxylin and eosin (H&E), Gram, Gridley, and Grocott stains; the eumycetes are more easily observed with H&E, periodic acid-Schiff (PAS), and Grocott/methenamine-silver stains (GMS).[8, 46]

Mycetoma[46] is endemic in tropical, subtropical, and temperate regions. Mexico, Venezuela, Sudan, India, Pakistan, Senegal, and Somalia have the highest occurrence rate. It is reported less frequently in the United States and other Latin American countries. Approximately 60% of cases of mycetomas are of actinomycotic origin. Mycetoma occurs most commonly in people who work in rural areas where they are exposed to acacia trees or cactus thorns. However, the disease has also been found in individuals living and working in the cities. Untreated mycetoma can affect the underlying bones, joints, or adjacent organs.[8, 46]

Image set 13.3

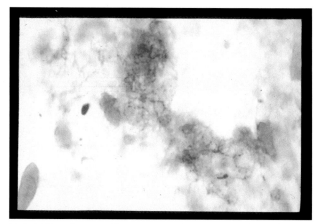

13.3-A Foot Lesion: Gram x1000 Gram positive branching bacilli *Actinomadura*

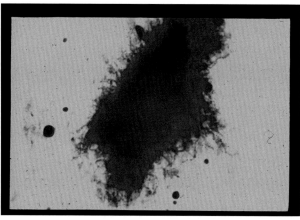

13.3-B Foot Lesion: Gram (culture) x1000 densely packed Gram positive branching bacilli *Actinomadura*

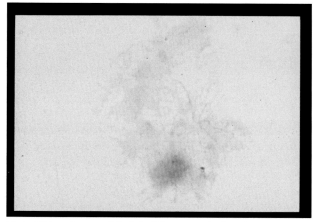

13.3-C Foot Lesion: Modified Kinyoun (culture) x1000 negative for partial acid fast

13.3-D Foot lesion: Ziehl-Neelsen (culture) x1000 negative for acid-fastness

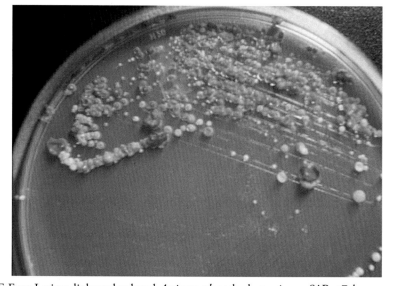

13.3-E Foot Lesion: light red colored *Actinomadura* slowly growing on SAB at 7 day

13.3-F Foot Lesion: coral red pigment produced by *Actinomadura* on SAB at day 14

13.4 CASE 56 – *MYCOBACTERIUM ABSCESSUS*

A cornea specimen was received for C&S, parasite, and fungal investigation. The Gram stain and fungal smear were negative for fungus and bacteria. A Giemsa stain was performed to rule out *Acanthamoeba* and turned out to be negative. However, many thin, non-branching, unstained filamentous forms (< 1μm) in bacillary morphology suspected of mycobacteria were seen[24] (*A: Image set 13.4*). Upon reviewing the Gram, tracks of unstained and vacuolated filaments suspected of mycobacteria were seen (*B: Image set 13.4*). The Gram-stained smear was overstained by ZN, and the previously unstained bacilli were now positive for acid-fastness. The specimen was forwarded to mycobacterial lab and culture confirmed *Mycobacterium abscessus*.

Image set 13.4

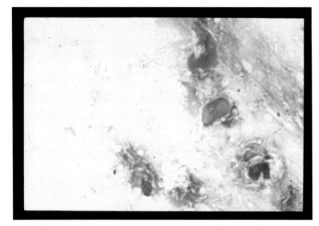

13.4-A Cornea: Giemsa x1000 unstained, curved, non-branching & filamentous bacilli (suspected mycobacteria)

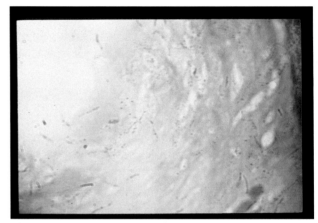

13.4-B Cornea: Gram x1000 Gram positive, beaded, unstained, curved, non-branching & filamentous bacilli (suspected mycobacteria)

13.5 CASE 57 – *MYCOBACTERIUM TUBERCULOSIS* (MTB)

A lymph node specimen from a middle-aged patient was received by microbiology for C&S and fungal infection. Pus cells were seen, but no bacteria were seen in the Gram-stained smear. However, many unstained filamentous forms were seen, indicating that organisms failed to take up Gram's reagents (*A: Image set 13.5*). The bacillary filamentous non-branching forms seen were suspected as mycobacteria. A fresh specimen smear was prepared and stained by ZN and observed under the microscope. Excellent red, slightly curved bacilli with attenuated ends positive for acid fast were seen in a ZN smear (*B: Image set 13.5*). The specimen was referred to the mycobacterial laboratory, which confirmed MTB by PCR.

The smear reader must keep a vigilant eye on the structures that are unstained and do not fit into the usual Gram positive or negative staining morphology. In these events the smear reader must think beyond bacteria when reading Gram smears during the search for bacteria in clinically important specimens. Any microorganism in tissue specimens is extremely important to identify in order to establish accurate clinical diagnosis so that appropriate therapy may be initiated.

Image set 13.5

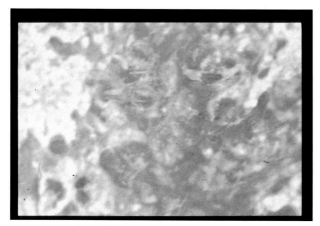

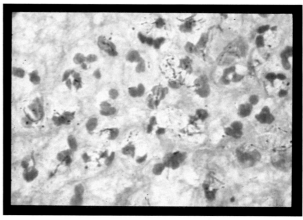

13.5-A Lymph Node: Gram x1000 unstained filamentous bacilli suggestive of mycobacteria

13.5-B Lymph Node: ZN x1000 Acid fast bacilli (MTB was confirmed by the reference lab)

CHAPTER 14

14.1 RARE FUNGI

The following fungi are rarely encountered in clinical microbiology (mycology). Although we do not have Gram smear images or any information as to how these organisms would behave when stained by Gram's method, we have some images of these fungi stained by specialized staining procedure such as GMS. Some day someone will encounter these organisms in a Gram-stained smear and expand on this part in a similar textbook with updated information.

See Image set 14.1 for the following rare fungi:

14.2 *HISTOPLASMA DUBOISII*

Histoplasma duboisii is a large variety of *Histoplasma capsulatum var. duboisii*. It is also known as African histoplasmosis. A limited number of cases have been reported for African histoplasmosis. The organism is acquired by inhalation and causes pulmonary lesions that may also spread to skin, bone, and lymph nodes usually located intracellular. The large cell size of *Histoplasma duboisii* (8–15 μm) must be differentiated from the large yeast cells of *Blastomyces dermatitidis* and *Loboa loboi*.[8]

14.3 *PARACOCCIDIOIDES BRASILIENSIS*

Paracoccidioidomycosis is also known as South American blastomycosis caused by single dimorphic species *Paracoccioides brasiliensis,* which is endemic in tropical and subtropical regions of Latin America.[8] The illness is associated with chronic and progressive respiratory condition lasting many years. Extrathoracic dissemination occurs in about 60% of patients spreading to other body sites.[8] *Paracoccioides brasiliensis* produces unique characteristics by producing large thick-walled double-countered round cells attached with daughter cells, giving the appearance of a marine wheel. Care must be taken to distinguish single daughter cells when they detach from the mother cell. The cell size ranges from μm 4 to 60 ìm (averaging 5–30 μm). The larger cell is thick-walled (1 μm) and doubly countered.[8] Multiple-budding cells attached to the mother cell provide excellent structures for easy detection and identification during direct microscopic examination of the clinical specimen processed by GMS stain.

14.4 ADIACONIDIA

Adiaspiromycosis is caused by a fungus, a self limiting pulmonary infection, prevalent in rodents in temperate climate around the globe.[8] The etiological agent causing adiaspiromycosis belongs to genus *Emmonsia* having two varieties in one species of *Chrysosporium,* such as *Chrysosporium parvum var parvum* and *Chrysosporium parvum var crescens.*[8] The solitary granuloma and disseminated pulmonary form are the two kinds of human adiaspiromycosis that have been seen.[8] *Chrysosporium parvum var crescens* is dimorphic fungus. Its conidia (2–4 µm) when inhaled begin to enlarge in the human body or at laboratory incubating temperature 37⁰C, becoming round and thick-walled adiaconidia reaching 200 to 400 µm (may reach to 700 µm).[8] The mycelium phase of this fungus grows at 20 to 30⁰C and does not appear in human tissue. Adiaconidia is seen in H&E preparation. Upon maturity, adiaconidia are refractile and have very thick walls measuring 20 to 70 µm. The interior of the adiaconidium is empty and hyaline in nature. The thick wall of adiaconidia stains black with GMS.

Image set 14.1

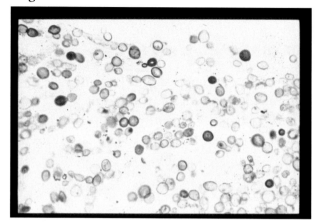

14.1-A Bone: GMS x400 *Histoplasma dubosoi*

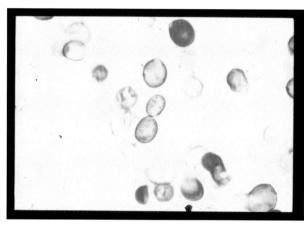

14.1-B Bone: GMS x1000 *Histoplasma dubosoi.*

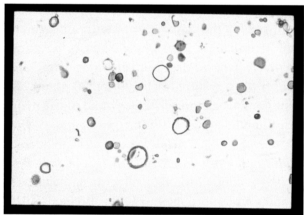

14.1-C Adrenal Gland: GMS x400 *Paracoccidioides brasiliensis* (thick-walled round cells, multiple buds resembling marine wheel)

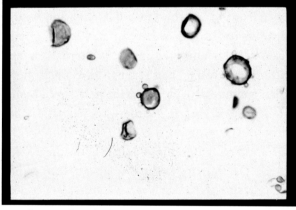

14.1-D Adrenal Gland: GMS x1000 *Paracoccidioides brasiliensis* (thick-walled round cells, multiple buds resembling marine wheel)

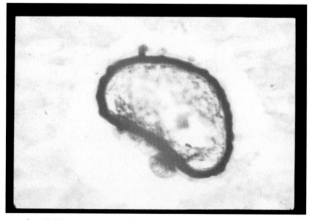

14.1-E Hamster Testicle: GMS x1000 Adiaconidium

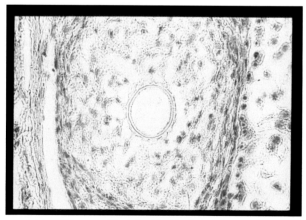

14.1-F Hamster Testicle: H&E x400 Adiaconidium.

CHAPTER 15

15.1 FUNGI IDENTIFIED BY DIRECT MICROSCOPY ONLY

The following fungi (*Rhinosporidium*, *Loboa*, and *Pneumocystis*) are recognized and identified by direct microscopy only. These fungi do not grow on the synthetic laboratory medium used routinely in clinical microbiology. Site specificity and clinical data of the patient helps to correctly identify these organisms when examined under the microscope. Different stains have been used to display structures of the organisms that are identified by direct microscopy only.

15.2 *RHINOSPORIDIUM SEEBERI*[8]

Rhinosporidium seeberi is a fungus in Zygomycetes usually causing infections of the superficial mucocutaneous sites such as the nasal and maxillary area and less commonly involving the conjunctiva, larynx, and genitalia. The infection usually occurs in males. It is hyperendemic in India, Sri Lanka, and Southeast Asia and occasionally encountered in the Western Hemisphere.[8] *R. seeberi* has two developmental stages, sporangia (mature) and trophocytes (immature).[8] The spherical sporangia when mature reach 100 to 200 μm or more in diameter and are often present in the mucosal polyps in numerous quantity.[8] The cell wall of the mature sporangia of *R. seeberi* is around 3 to 5 μm thick. Mature sporangia contain numerous round and oval endospores ranging from 2 to 10 μm. The morphology of *R. seeberi* is well displayed in H&E preparation. GMS and PAS stain sporangia and endospores deeply but not trophocytes. Gram stain images (*A & B: Image set 15.2*) are showing some important features of *R. seeberi* close enough to compare with similar-looking structures such as *Coccidioides*. All other structures of large and round morphology should also be compared with *R. seeberi*. The thick cell wall and larger endospores and the much larger size of *R. seeberi* sporangium differentiate it from the *Coccidioides* spherule, which has a comparatively thinner cell wall and small-sized endospores (2 to 3 μm). The structures seen in Gram-stained smears resembling *R. seeberi* are compared with *Coccidioides* and ruled out based on the size and specimen site. When confirming structures seen as *R. seeberi* it must be confirmed by special stain such as H&E and GMS to fully visualize the internal and external details of these organisms. The H&E image (*C: Image set 15.2*) shows characteristic sporangia, while the GMS image (*D: Image set 15.2*) stains the sporangium deeply with unstained endospores. It is difficult to tell the difference between *R. seeberi* and *Coccidioides*. The CW image (*E: Image set 15.2*) does not provide internal details, and the sheer large size of the sporangium is not sufficient to identify *R. seeberi* based on visually brighter sporangium. The KOH wet prep image (*F: Image set 15.2*) is somewhat more promising than all other staining procedures for its simplicity and ease to make a prep, and all the details that need to be seen stand out clearly, such as thick cell wall and larger-sized endospores. The site specificity would presumptively identify this organism as *R. seeberi* pending confirmation by H&E and other specialized techniques such as molecular studies.

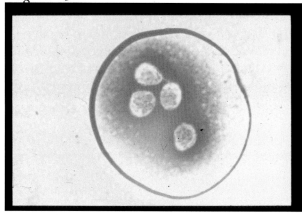

15.2-A Nasal Biopsy: Gram x1000 endospores of
Rhinosporidium seeberi

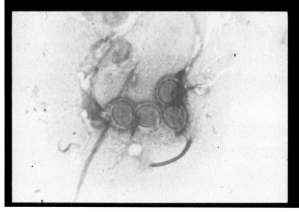

15.2-B Nasal Biopsy: Gram x1000 endospores of
Rhinosporidium seeberi

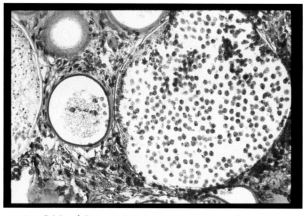

15.2-C Nasal Biopsy: H&E x250 mature and immature
sporangia of *Rhinosporidium seeberi*

15.2-D Nasal Biopsy: GMS x100 mature sporangium of
Rhinosporidium seeberi (resembling *Coccidioides*)

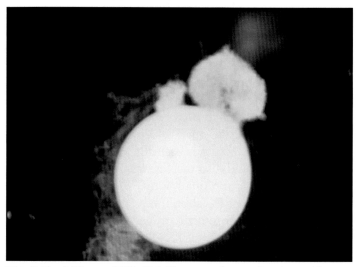

15.2-E Nasal Biopsy: CW x100 *Rhinosporidium seeberi* (resembling
Coccidioides); internal details are not visible

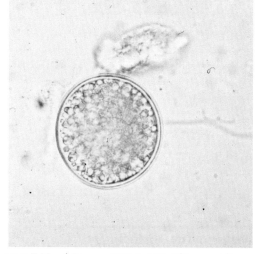

15.2-F Nasal Biopsy: KOH x100 *Rhinosporidium
seeberi* (resembling *Coccidioides*); internal details
are clearly visible

15.3 *LOBOA LOBOI*[8, 9, 33]

Lobomycosis is a cutaneous infection starting as indurated nodules usually involving the feet, legs, buttocks, or face. Older lesions often turn verrucous and ulcerate. *Loboa loboi* is restricted to South America and parts of Central America. Fungal cells are abundant in cutaneous lesions and are detected by H&E and GMS as round and oval cells in chains (usually three to eight cells). Budding form of yeast-like cells ranging 5 to 12 μm (average 8 μm) in diameter are attached through a narrow tube-like structure called an isthmus.[8] The cell wall is doubly countered and about 1 μm in thickness (*A & B: Image set 15.3*).[8]

Image set 15.3

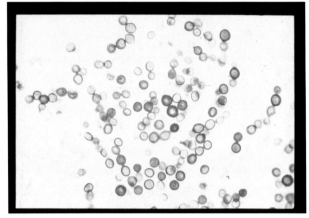

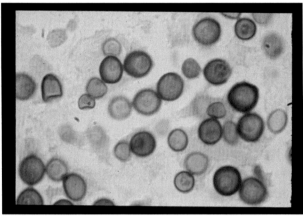

15.3-A Skin Biopsy: GMS x400 *Loboa loboi* (yeast-like cells in chains connected by a tube like structure)

15.3-B Skin Biopsy: GMS x1000 *Loboa loboi* (yeast-like cells in chains connected by a tube like structure)

15.4 *PNEUMOCYSTIS JIROVECII*[5, 9, 29]

Pneumocystis jirovecii is formerly known as *Pneumocystis carinii* and found in the interstitial fluid of immunocompromised hosts, especially those who are HIV positive. A BAL specimen is the superior specimen over sputum to recover *Pneumocystis* cysts (PC) clustered in foamy exudate. Various methods such as GMS, CW, Giemsa, IFA (antigen detection), and molecular techniques have been in use to detect PC. All direct-staining procedures have advantages and disadvantages. GMS and CW detect cyst forms and are highly sensitive procedure. CW stain is nonspecific, but PC detection is comparable. Both the GMS and CW lack specificity, and false negative results may be possible. IFA is extremely sensitive; therefore, the risk of false positive occurrence would require another procedure for comparison and confirmation during doubtful circumstances. Fungi-Fluor (CW) stained smear made from a BAL specimen shows excellent intensely stained PC surrounded by a faint cyst wall. CW detection is achieved by observing morphological characteristics that do not show such structures in any other fungus. PC are 3 to 7 μm in diameter, mostly round, with centrally located internal parenthesis like "dots." The detection is basically made by morphology (*A: Image set 15.4*). A Gram-stained smear processed for C&S would also display suspicious structures for PC. Round, hollow, and clustered honeycomb material is the indication of a foamy exudate (*B & C: Image set 15.4*). The cyst forms are sometimes more obvious, and suspicion of PC should be clarified by staining the smear with GMS (*D: Image set 15.4*) or direct detection of immunofluorescent antigen (IFA) by monoclonal antibodies (*E: Image set 15.4*). Wet prep such as unstained KOH (*F: Image set 15.4*) may also be able to show foamy exudate containing PC. The clear detection is not possible by this method, but there is a suspicion of specimen material with possibility of PC and foamy exudate. During such situations the smear should be stained by CW, GMS, or IFA to confirm the presence of PC suspected in Gram and KOH smears.

Image set 15.4

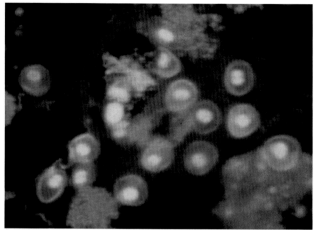

15.4-A BAL: FS x1000 *Pneumocystis jiroveci* (*carinii*). Observe intensely staining internal dots surrounded by faintly staining cyst wall

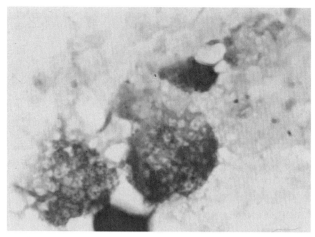

15.4-B BAL: Gram x1000 *Pneumocystis jiroveci* (*carinii*). Observe honey-comb appearance and red dot in the center of the cyst

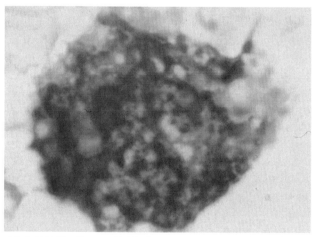

15.4-C BAL: Gram x1000 *Pneumocystis jiroveci* (*carinii*). More cysts with central dots are visible

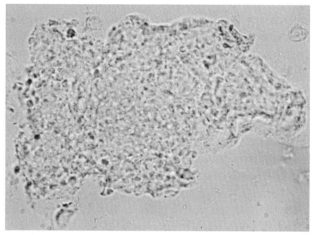

15.4-D BAL: KOH x400 *Pneumocystis jiroveci* (*carinii*). Observe large area of foamy exudate and PC cysts giving appearance of honey-comb

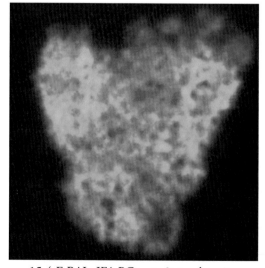

15.4-E BAL: IFA PC cysts & trophozoites

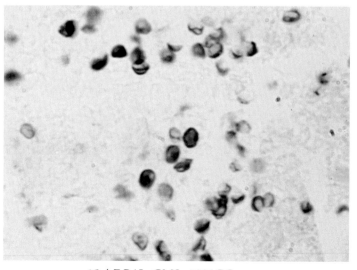

15.4-F BAL: GMS x1000 PC cysts.

CHAPTER 16

The following are a few cases that did not have fungus in the corresponding clinical specimens. These specimens were negative for C&S as well as fungus. However, the specimen smears processed for C&S and fungus revealed organisms in direct smear that needed attention for the appropriate therapy to the patient. In some cases patient data was required to make an appropriate decision. The specimen site was equally important to resolve certain case scenarios. It is for this reason that, during Gram smear examination, microbiology must be kept in mind by looking beyond bacteriology.

16.1 *PROTOTHECA*[8, 20, 23, 33]

Prototheca is achloric algae. *Prototheca wickerhamii* and *Prototheca zopfii* are pathogenic species found worldwide, but they rarely cause infections among humans. *Prototheca* is isolated from patients' specimens from time to time and most often confused with yeast due to the morphology of the organism as being pale moist and yeast-like. *Prototheca* grows in synthetic laboratory medium at 25⁰C and 37⁰C. Unique microscopic structures and a commercial product such as API 20C AUX identify *Prototheca* to the species level. *Prototheca* causes two types of infections: cutaneous and olecranon bursitis. Cutaneous infections occur in two-thirds of the cases, progressing to verrucous or scaling lesions on hands, legs, and feet.[8] Such infections occur in immunocompromised hosts or debilitated patients. Olecranon bursitis produces subcutaneous nodules adjacent to the elbow and caused by trauma in healthy individuals.[9] *Prototheca* species differ in size; small-size *P. wickerhamii* ranges from 2 to 12 μm in diameter, and *P. zopfii* ranges from 10 to 25 μm and has vacuolated cytoplasm.[8] *Prototheca* produces sporangia containing up to 20 endospores (only four to eight are visible). The cell wall of *Prototheca* is stained by GMS and PAS, showing the morula form of *P. wickerhamii* (infrequent with *P. zopfii*), caused by internal cleavage and continued division until the formation of endospores.[8] In routine microbiology, *Prototheca* may be easily visualized by wet prep (KOH) or iodine staining.

16.2 CASE 58

A 57-year-old female patient with a long history of rheumatoid arthritis developed chronic abdominal pain as a complication of her immunosuppressive therapy. Peritoneal fluid was sent to the microbiology lab for C&S & fungal investigation. No PMN or bacteria was seen in C&S, and no fungal elements seen in a fungal-stained smear. Two weeks later two colonies of white mold were seen on mycology culture media and identified as *Histoplasma capsulatum*. Recovery of *Histoplasma capsulatum* from peritoneal fluid was a surprise for the clinician. The patient was treated with antifungal agent and discharged. Three months later the patient was re-admitted to the hospital for olecranon bursitis. Synovial fluid from her left elbow was sent to the microbiology lab for *Histoplasma* investigation. A direct-specimen smear stained by CW was negative for *Histoplasma*, but there were several round, oval and nonbudding cells seen in variable sizes. Some cells were mature, with internal septation and endospores depicting excellent morphology of *Prototheca*. The diagnosis was made based on direct smear results only; culture remained negative for six weeks and was discarded. No growth was reported after the six-week incubation.

The IDS team found this case challenging for diagnosis of unusual organisms in unexpected situations, suggesting that the laboratory aspects of this case highlight the need for communication between consultants and the microbiology laboratory and the utility of direct microscopy in diagnosis (*A to D: Image set 16.2*).

Image set 16.2

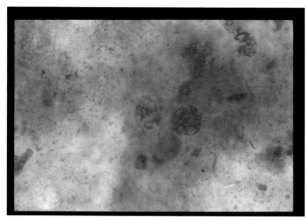

16.2-A Synovial Fluid: Gram x1000 round non-budding cells of *Prototheca*

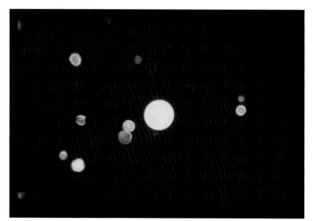

16.2-B Synovial Fluid: CW x400 immature sporangium and few small round to oval endospores of *Prototheca*

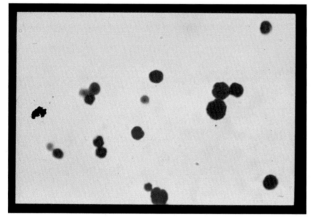

16.2-C Synovial Fluid: Gram (made from culture) x1000 round endospores and immature sporangia of *Prototheca*

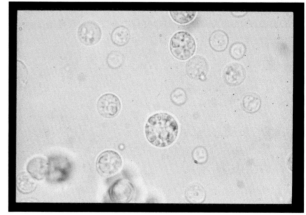

16.2-D Synovial Fluid: KOH x1000 mature and immature sporangia of *Prototheca*

16.3 CASE 59 – *ACANTHAMOEBA*

It has been suggested that structures resembling yeast must not be called yeast unless budding forms have been demonstrated at some point within the same smear (it does not necessarily have to be the same microscopic field). In case no budding form is seen in the entire smear, do not report structures as yeast, but define their morphology explicitly as they appear under the microscope, such as the shape, size, and any other identifiable details seen under the microscope.

Cornea specimens were received from the eye clinic for C&S and fungal investigation. The Gram smear showed pus cells but no bacteria seen in some cases while in other cases yeast cells were reported, since structures seen under the microscope were oval/spherical bodies between 6 and 10 μm in diameter (*A: Image set 16.3*). In mycology, a fungal stain using CW showed round and oval non-budding forms in clusters (*B: Image set 16.3*). The site of the specimen was a help to suspect structures as cysts of *Acanthamoeba* and not yeast. No bacteria or fungus was isolated from the specimen. The culture was set up on a nutrient plate overlaid with *E. coli* and incubated at 37⁰C. A Giemsa smear was prepared from the culture and examined under the microscope and displayed numerous polygonal cells consistent with *Acanthamoeba*[32] (*C: Image set 16.3*) with clear separation of cyst wall and internal membrane. A Gram smear prepared from the culture also displayed polygonal structure but was less conspicuous than the Giemsa image (*D: Image set 16.3*).

Image set 16.3

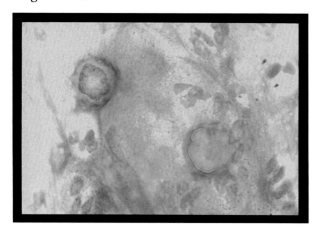

16.3-A Cornea: Gram x1000 round, thick-walled or double walled non-budding cells of *Acanthamoeba* cysts (note polygonal cell wall)

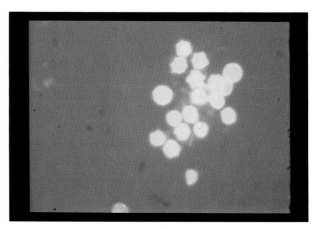

16.3-B Cornea: CW x400 non-budding cells of *Acanthamoeba* cysts (note polygonal cell wall)

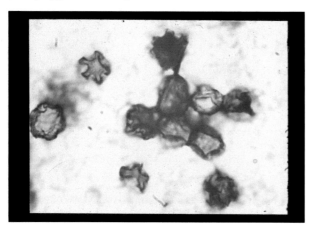

16.3-C Cornea: Giemsa x1000 round, thick-walled or double walled non-budding cells of *Acanthamoeba* cysts (note polygonal cell wall)

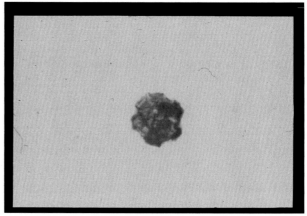

16.3-D Cornea: Gram (made from culture) x1000 round, polygonal, non-budding cell of *Acanthamoeba* cyst

16.4 CASE 60 – *STRONGYLOIDES*

A BAL specimen from a middle-aged lady suffering from lupus and other immunological conditions was received for C&S and fungal investigation. The Gram smear was reported out as negative for bacteria. A CW done in mycology showed some filamentous organism that did not fit into the morphology of bacteria or fungus. The organisms seen were much bigger than the any known fungus and were identified as larvae of parasites (*B: Image set 16.4*). The Gram smear was reviewed, and huge larvae of *Strongyloides stercoralis*[32] were seen and confirmed by parasitology (*A: Image set 16.4*).

Image set 16.4

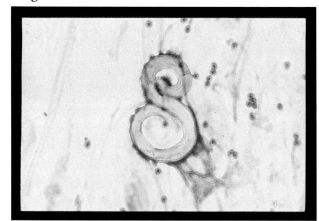

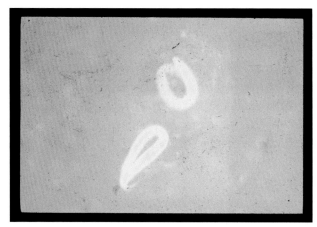

16.4-A BAL: Gram x250 *Strongyloides stercoralis* (larva)

16.4-B BAL: CW x200 *Strongyloides stercoralis* (larva).

16.5 CASE 61 – MICROSPORIDIA

A conjunctiva specimen was received for C&S and fungal investigation from a male (HIV-positive) host. The Gram smear showed pus cells and also Gram positive organisms interpreted as Gram positive bacilli (*A: Image set 16.5*). A CW smear in mycology also showed these organisms brightly fluorescing under the UV light microscope (*B: Image set 16.5*). The fact about CW stain is that it brightens up the organisms by binding to the chitin and cellulose in the cell wall of fungi and parasites. Bacteria do not have chitin or cellulose in their cell walls; as a result they do not keep the intensity up for long. This alerted us to review the Gram smear and observe that the organisms appearing as Gram positive coccobacilli may not belong to bacterial entity after all. None were found budding in the Gram or CW smear. The organisms were suspected as microsporidia,[27] which was confirmed by parasitology.

Image set 16.5

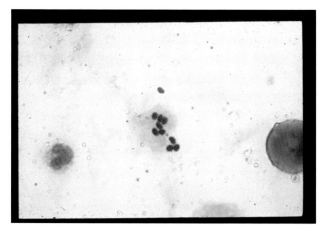

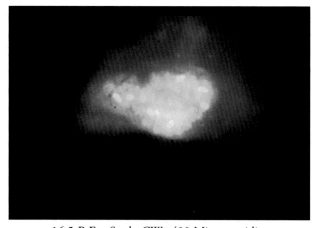

16.5-A Eye Swab: Gram x1000 Microsporidia

16.5-B Eye Swab: CW x400 Microsporidia.

16.6 MYOSPHERULE[8, 9, 19, 25, 28, 33, 44, 47]

Myospherule structures are made up of altered erythrocytes. Myospherule is divided into two categories: 1) iatrogenic spread (causality with petroleum-based ointment and gauze packs) seen in the paranasal cavity, ocular cavity, intracranially, and in dermal tissue; and 2) those cases where erythrocytes are altered by endogenous fat or lipids. The myospherules appear as round bags of spherules surrounded by a thin, refractile membrane when examined under bright-field microscopy. The spherules are closely and irregularly packed together and do not show any internal structure; however, in some cases, small granules or irregular blobs of eosinophilic material may be seen. The approximate size of the spherule ranges from 10 to 15 μm in diameter. Myospherules do not stain by GMS and PAS. However, they are visualized by H&E and are suspected to be *Rhinosporidium* or *Coccidioides*. In H&E prep, some spherules show thin, pale, eosinophilic membrane and others a brown, thicker, and coarser membrane, with irregular granularity on the outside. There is no information available at this time as to how myospherules would behave when stained by Gram's method. (See *Image set 16.6*.)

Image set 16.6

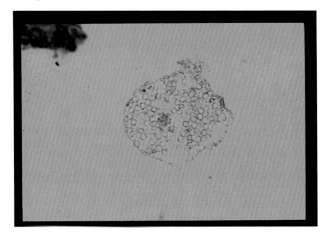

16.6-A Skin Biopsy: H&E x400 Myospherule

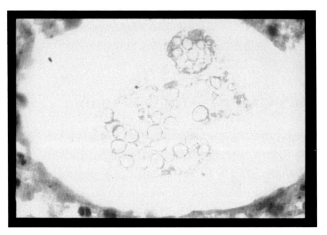

16.6-B Skin Biopsy: H&E x1000 Myospherule.

CHAPTER 17

17.1 PATHOLOGY VERSUS ROUTINE MICROBIOLOGY

Clinical specimens sent to microbiology for C&S, fungal, viral, and mycobacterial analysis are also sent to the histopathology for pathological studies to establish a diagnosis of patients' clinical conditions. The pathology lab processes clinical specimens using various specialized staining procedures such as GMS, PAS, H&E, FM, and mucicarmine and applies other techniques necessary to identify etiological agent causing infection. Clinical microbiology, on the other hand, runs simple, rapid, cost-effective, and routine procedures to identify the infectious agents. The majority of times routine microbiology stains are sufficient in terms of detecting microorganisms. However, there are some difficulties when microbiology is unable to control certain variables that cause the direct smears to turn falsely negative. The major difficulty that microbiology comes across is in receiving insufficient quantities of the specimen as well as the site of the specimen not containing microorganisms. Pathology, on the other hand, processes specimens in wax blocks, cuts sections in single layers, and embeds the entire specimen on the microscopic slide without the fear of losing the material from the glass slide during the staining and washing steps. This is not possible with routine microbiology staining procedures. Microbiology smears frequently miss organisms as compared to pathology, which seldom comes across such a scenario. Comparatively, pathology slides are superior in contrast and clarity as compared to the routinely stained smears in microbiology due to artifact and reagent deposits. However, the pathology labs often run into problems in terms of identifying structures that they have seen in their GMS or other smears in greater quantity. At this point they seek microbiology's help for interpretation of the organisms. This type of mutual communication between the two departments is most useful for the patient diagnosis as well as providing learning opportunities that usually come from the problematic cases. Sometimes microbiology visualizes structures in low numbers and not enough to make a decision. At this point, pathology-stained smears are visualized by microbiology to confirm diagnosis of the patient's clinical condition suspected by microbiology-stained techniques. The following are a few examples to address the issue surrounding the staining specificity and smear reader's aptitude as well as the factors affecting the precise understanding in terms of recognition of the structures that should not pose a problem in normal situations.

17.2 CASE 62

A male patient was diagnosed with cancer. His nasal-maxillary biopsy specimen was sent for histological studies. The pathology lab reported filamentous fungus ball and round cells in the morphology of *Aspergillus* and *Cryptococcus* seen in GMS (*A: Image set 17.2*) and PAS (*B: Image set 17.2*) smears. Histology smears were reviewed in mycology, and no *Cryptococcus* or fungus ball involving *Aspergillus* were observed, but they contained variable shaped and sized fungal elements in yeast and pseudohyphae morphology. Pathology was requested to prepare Gram and melanin smears to be viewed by a mycology expert. The melanin (FM) stained smear was more clear than the GMS and PSA smears. Round cells in budding and non-budding forms as well as pseudohyphae with constrictions at segments and some wider hyphae with clear septum were seen (*C: Image set 17.2*). No pigmented structures were seen. The Gram smear showed some elongated curved cells with pointed ends mimicking yeast (*D: Image set 17.2*). Pseudohyphae were also observed in other microscopic fields. The structures seen resembled *Candida*, although the newer and older fungal elements behaved in different ways, producing more than one distinct morphology of the single organism. The specimen was never submitted to the microbiology lab for culture. The specimen was sent to the reference lab for molecular testing. *Candida albicans* was identified by DNA studies.

Image set 17.2

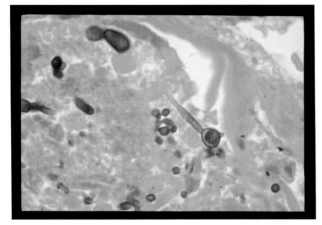

17.2-A Nasal Mass: GMS x1000 yeast with pseudohyphae (pathologist reported as *Aspergillus* and *Cryptococcus*)

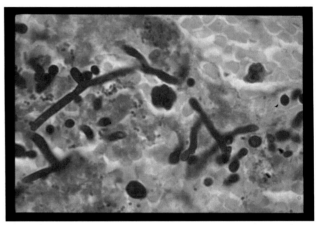

17.2-B Nasal Mass: PAS x1000 yeast with pseudohyphae (pathologist reported as *Aspergillus* and Cryptococcus)

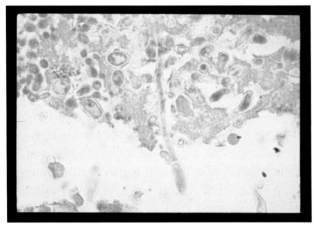

17.2-C Nasal Mass: FM x1000 yeast with pseudohyphae (no melanin observed in the cell walls of yeast)

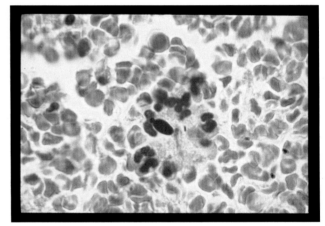

17.2-D Nasal Mass: Gram x1000 yeast seen (pathologist reported as *Aspergillus* and *Cryptococcus*)

17.3 CASE 63

The pathology lab reported *Mucor* from a lung-transplant patient's BAL specimen. Septate hyphae in BAL were reported as *Mucor* by pathology seen in GMS (*A: Image set 17.3*), PAS (*B: Image set 17.3*) and other smears. Mycology also had a specimen that showed septate hyphae and recovered *Aspergillus flavus* from the corresponding culture and reported it to the clinician. Within the next 24 to 48 hours the clinician called the mycology lab and asked for clarification just in case *Mucor* in the specimen were missed that were reported by the pathology lab. GMS and PAS smears were reviewed and found to have no *Mucor*-like fungal elements but parallel-walled septate hyphae in them. The Gram was also retrieved and reviewed, and there were found much wider true hyphae with frequent septation as well as parallel walls (*C: Image set 17.3*). Initially, Gram smear results did not indicate fungal elements in the report. The Gram smear was reviewed upon seeing massive amount of aseptate hyphae in the CW (*D: Image set 17.3*). Confirmation about septate hyphae and the culture identity *Aspergillus flavus* was sent to the clinician as well as to the pathology lab. The pathology lab issued an amended report to maintain accuracy and the consistency of the results reported to the clinician.

Image set 17.3

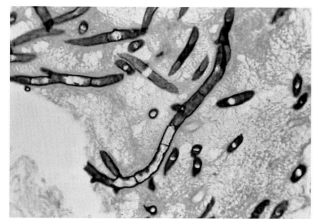

17.3-A Lung Biopsy: GMS x400 septate hyphae (pathology report was sent out as *Mucor*)

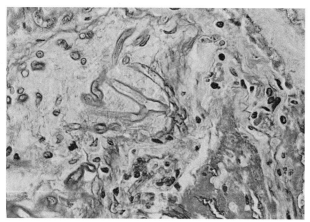

17.3-B Lung Biopsy: PAS x400 septate hyphae (pathology report was sent out as *Mucor*)

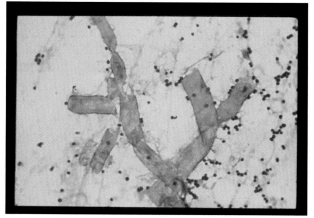

17.3-C Lung Biopsy: Gram x1000 septate hyphae (pathology report was sent out as Mucor)

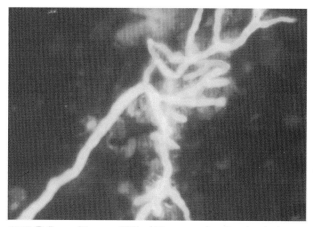

17.3-D Lung Biopsy: CW x400 septate hyphae (pathology report was sent out as Mucor)

17.4 CASE 64

A lung tissue was submitted to histopathology for *Histoplasma* investigation. The pathology lab saw some structures and had no idea how to interpret them. Mycology was consulted, and a pathology smear stain by GMS (*A: Image set 17.4*) was reviewed. *Histoplasma* was ruled out, but the structures seen in the GMS smear did not fully reveal the identity of the fungal elements seen. The structures were overlapping between fungal hyphae belonging to dark fungus, since toruloid structures were seen. Annellidic types of structures were also suspected that are not routinely seen in direct smears. Since the specimen was left in formalin, mycology tried to make a wet prep (*B: Image set 17.4*) and also stained by Gram's method (*C: Image set 17.4*) and CW (*D: Image set 17.4*).

Structures seen in all the smears above did not resolve the conflict. At some point it appeared as pseudohyphae; then again it looked like a dark fungus behaving in an unusual way. The specimen could not be cultured since it was placed in formalin for histopathology studies. The specimen was sent to the reference lab for molecular studies. DNA studies identified the fungus as *Candida albicans*.

Microscopic observations of the fungal elements seen were similar to histology findings. Numerous branched hyphae with frequent septum, vase-shaped, swollen cells, and elongated hyphal fragments mimicking annellidic types of structures were observed. At this point it was very hard to categorize these structures as septate hyphae, pseudohyphae, or any other type of morphology. In the absence of culture, the precise interpretation of the structures to give it a definitive name was difficult. It was to everyone's surprise to find that the DNA studies identified the fungal elements seen in the direct smear as *Candida albicans*.

Image set 17.4

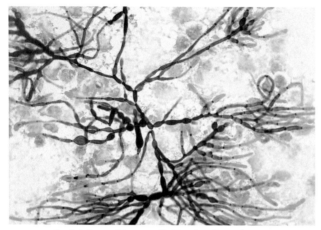

17.4-A Lung Tissue: GMS x400 branching pseudohyphae like structure

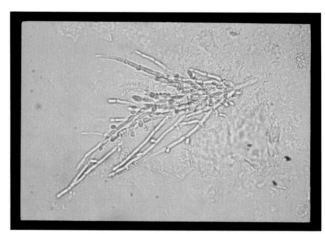

17.4-B Lung Tissue: WP x400 branching pseudohyphae like structure

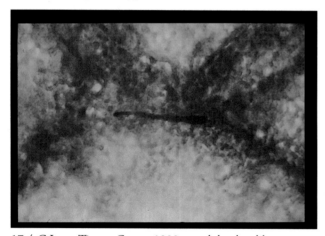

17.4-C Lung Tissue: Gram x1000 pseudohyphae like structure

17.4-D Lung Tissue: CW x400 branching pseudohyphae like structure

CHAPTER 18

18.1 MISCELLANEOUS

18.2 HELPING HINTS FOR THE MICROSCOPIC APPEARANCE OF FUNGAL ELEMENTS

Depending on the specimen, slides stained by CW may show fungal elements as brilliant yellow-green fluorescing yeast cells, septate, aseptate hyphae, or pseudohyphae.

Generally, yeast cells may be presumptively identified as *Candida*, *Blastomyces*, and *Cryptococcus* species based on specific morphological features.

Candida shows oval yeast cells with pseudohyphae. Yeasts other than *Candida* may show oval or round cells in various sizes.

Blastomyces appears in its yeast phase in tissues as large, thick-walled, budding yeast with a broad base, giving the appearance of a "figure 8," and has an average size of 8 to 15 μm (range 3 to 30 μm).

Coccidioides shows spherules with or without endospores typically ranging from 10 to 100 μm but may reach to 200 μm. The endospores of *C. immitis* may not fluoresce as intensely as other fungal elements would. Immature spherules of *C. immitis* can resemble *Blastomyces* if the two spherules are side by side. However, unlike *Blastomyces*, the endospores and spherules of *C. immitis* do not bud.

Cryptococcus shows round, budding cells that are 2 to 20 μm with a narrow base at the attachment site. The observation of the capsule is indicative of *Cryptococcus*. The capsule of *Cryptococcus* does not stain with the CW and may appear as a black halo surrounding cryptococci. Using KOH or wet prep, observe the darkness of the cell wall.

Histoplasma shows small, oval yeast cells with a narrow isthmus, 2 to 4 μm in diameter, and is often found intracellular in macrophages. CW is not suitable to identify *Histoplasma* by morphology. However, a Gram-stained smear usually displays characteristic intracellular structures of *Histoplasma* stained pink or remaining unstained and appearing as hollow or vacuoles.

Malassezia species show very small yeast cells with unipolar budding (2 to 5 μm). When present on skin as normal flora, *Malassezia* cells are observed as oval or round and may appear like "bottleneck" or "bowling-pin" morphology under the microscope. However, when it turns into an infectious state, *Malassezia* changes its morphology to round cells and short filamentous forms like "meatballs and spaghetti."

Septate hyphae with a width of 3 to 6 μm may show dichotomous branches. Forked type of branching is suggestive of *Aspergillus* species and other opportunistic fungi, such as *Fusarium* species, *Acremonium* species, or *Scedosporium*.

It is suggested that the pigment of the fungal elements must be observed (by examining the smear under bright field) to detect the dematiaceous group of fungi. Certain fungi in the dematiaceous group would produce structures whose morphology is worth paying attention to. A fluorescent stain such as CW would not be able to demonstrate melanin pigment in the fungal cell walls. However, fungal elements in the dematiaceous group produce typical morphology such as pseudohyphae or toruloid structures, which may appear as cells in chains mimicking yeast. Such structures give a clue to the nature of the dark fungus involved. Many times dark-walled fungal elements are observed in the Gram or wet prep (KOH). However, dark fungi do not always show pigment in their cell wall. In such circumstances, a special stain Fontana Masson is done to demonstrate melanin pigment in the cell walls of dematiaceous fungi.

Broad, non-septate hyphae, two to three times the width of *Aspergillus* hyphae may be suggestive of Zygomycetes (*Mucor* species, *Rhizopus* species, *and Absidia* species). Pay special attention to the irregular width such as too wide at one end and too narrow at the other end. The variation in width is due to the lack of septum within close vicinity, causing the cytoplasmic stream to fluctuate during physical or mechanical pressure applied to these vital structures. As a result, Zygomycetes are known to produce variable width. It must be noted that certain *Aspergilli* produce irregularly wide hyphae with sparse septation mimicking non-septate fungi. *Aspergillus flavus* has been seen to produce wide, ribbon-type hyphae, confusing it with aseptate hyphae.

18.3 COMMON REASONS FOR MISSING FUNGAL ELEMENTS IN GRAM SMEARS

The Gram stain is the most commonly used procedure in clinical microbiology for the detection of bacteria from clinical specimens. The Gram stain procedure is not typically considered acceptable to detect FE from clinical specimens. However, it has created a specified niche within direct microscopy to find organisms other than bacteria in Gram-stained smears. One can learn how to examine Gram smears for the detection of FE in clinical specimens. Listed below are common reasons why FE is missed in direct Gram smears.

1. The clinician fails to request fungal investigation on a clinical specimen

2. Technologists are not trained to identify fungi during Gram smear examination

3. Technologists lack experience in mycology

4. A small number of FE are present in the Gram smear

5. The use of the oil immersion lens to search for bacteria examines a very small microscopic field

6. FE present in the smear are distorted

7. FE present in the smear are ignored as specimen debris

8. Unstained FE present in the Gram smear appear without contrast

9. FE are masked by Gram stain reagents

10. Poor specimen, selected from an area not containing FE

11. Gram stain not an ideal procedure for fungal detection

CHAPTER 19

19.1 PROS AND CONS ABOUT SPECIAL (SPECIFIC) STAIN

- GMS does not stain cellular material

- Cellular material is essential to diagnose certain fungi

- H&E is most useful to suspect area in the specimen material indicating the host's response to the invading organisms

- Gram stain is useful to show cellular material and the organisms

GMS has a similar drawback as CW, since both highly sensitive test procedures fail to demonstrate host reaction. Therefore, the cellular response is missing that helps make a decision on those organisms to identify them, such as *Histoplasma capsulatum*. In a histology laboratory the pathologist always would want to recognize granuloma in H&E but may not be able to assess it clearly from GMS; however, the procedure is very specific for identification of fungi from the clinical specimens. In our case, we don't really need to examine the H&E smear for granuloma, which is also not a routine practice in a clinical microbiology lab. The Gram-stained smear alone provides enough information about cellular response such as PMNs for routine purposes that is equally credible.

19.2 FONTANA MASSON (FM) FOR MELANIN STAIN[8, 31]

It is suggested that chromaffin reaction oxidizes melanin or a melanin precursor as it reduces silver. Cell walls appear black and background pink. It is useful to stain cell walls of dematiaceous fungi when the nature of melanin pigment is not evident. It is useful for *Cryptococcus neoformans* (natural color of cell walls masked). However, care must be taken to correctly interpret melanin-positive cells and not call dark brown material in masking the fungal elements that may be present in severely necrotic specimen grossly appearing dark or reddish brown. The clue to separate false positive melanin stain is to observe the cell wall, which should stain brown with Fontana Masson (melanin stain), and not the grossly surrounded fungal elements by dark necrotic debris.

19.3 GOMORI METHENAMINE SILVER (GMS) STAIN[8, 31]

Chromic acid oxidizes polysaccharides in the cell wall to aldehydes, which in turn reduces methenamine silver nitrate to metallic silver and makes fungal cell walls look brown to black against the green background. RBCs and naked nuclei can mimic yeasts. No cellular information is available.

19.4 HEMATOXYLIN AND EOSIN (H&E) STAIN[8, 31]

Hematein-mordant stains DNA and nuclear proteins as blue, cytoplasm and other components as red, and RBCs bright red. They are excellent for displaying a cellular response to certain fungi and are able to find them in tissue.

19.5 MUCICARMINE STAIN (MC)[8, 31]

Mucicarmine stain detects the capsule of *Cryptococcus neoformans,* especially when other fungi in similar (usually round) morphology such as *Blastomyces dermatitidis* and *Rhinosporidium seeberi* are seen in direct smears. The fungi producing species-specific structures and the staining pattern would play a part in differentiation. *Cryptococcus neoformans* would stain the body of the cell and the capsule as pink.

CHAPTER 20
THE QUIZZES

The quizzes in this chapter are provided for the readers to practice. Consider these images as appearing under the microscope in your own microbiology laboratory. Examine the images in each quiz and make a decision about the microscopic structures shown. Decide how you would have handled each case independently. Each quiz is followed by an answer.

20.1 QUIZ 1

Examine two images: (A) from an ear and (B) from BAL. Both images are taken from a Gram stain at x1000 magnification. Interpret your findings to the clinician.

Image set Quiz 1

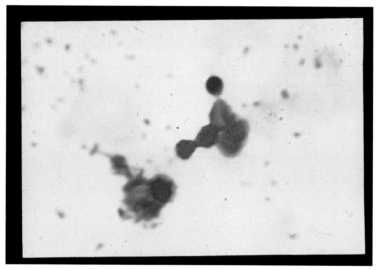

Quiz 1-A Ear: Gram x1000

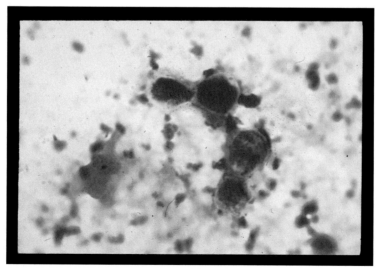

Quiz 1-B BAL: Gram x1000

ANSWER TO QUIZ 1

First of all, note the nature of the specimens—the ear swab and the BAL. Before starting to describe structures observed under the microscope, try to lay out some sort of strategy, such as the likelihood of the organisms expected to grow in reference to the site of the specimens. The ear swab sample is superficial in nature; i.e., not a part of the deep-seated infectious area. As far as mycology is concerned, *Aspergillus niger* is the most likely fungus to cause Otomycosis, although other fungi causing ear infection have also been documented.

Looking at image A: The structure seen under the microscope displays round to oval cells depicting budding forms. Yeast in the ear are usually seen as innocent bystanders or colonizers and may not be responsible for playing a role in the infectious process unless suggested otherwise. The next step to resolve the conflict is to determine what else this structure could be that produced infection in the patient's ear. The corresponding C&S report was negative. During such situations, the structures seen under the microscope must be thoroughly scrutinized before reporting direct smear results to the clinician. Care must be taken in interpreting direct smear results to avoid a corrective action later on. The round cells seen under the microscope are actually fungal conidia that have expanded to increase in size and are about to develop a germ tube to grow into mycelium form. The culture grew *Aspergillus niger*.

Image B comes from a BAL specimen. One would probably like to ask a question: Is the host immunocompromised? Once again, as in image A, this image is also showing round to oval cells in budding form. Something is different in this image—the size is larger and the cell wall appears thicker. A closer look would reveal that the budding pattern of the yeast-like cell is broad. Therefore the best-fitting statement about the structures seen would be "large round to oval, thick and double-walled budding yeast with broad base suggestive of *Blastomyces dermatitidis*", which was recovered later on from the corresponding culture and confirmed.

20.2 Quiz 2

Two specimens from two different patients were received: (A) heart valve and (B) sputum. Examine the structures seen under the microscope at x1000 magnification. How would you report the results to the clinician?

Image set Quiz 2

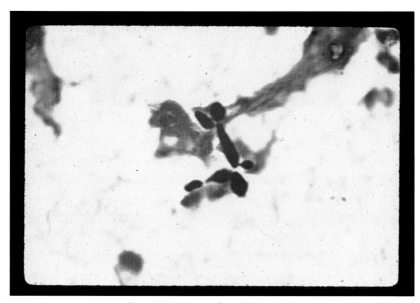

Quiz 2-A Heart valve: Gram x1000

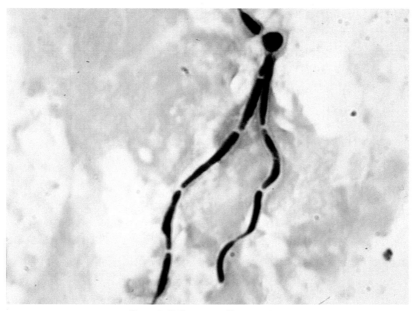

Quiz 2-B Sputum: Gram x1000

ANSWER TO QUIZ 2

The images come from two important specimens: A) the heart valve; and B) sputum.

It must be remembered that most fungal infections are acquired by inhalation. Therefore, the mediastinal is the chosen site for fungal infections. Fungi entering the human host after they are inhaled may either be eliminated by the immune system rapidly or stay and colonize the mucous membrane and surfaces of the respiratory tract (trachea, lung, and alveoli). Individuals whose immune system is impaired either temporarily or permanently may risk acquiring fungal infections more easily than people having their immune system intact. The degree of invasion and dissemination would depend on the organism involved, the patient's immune status, and their clinical conditions.

Fungus reaching the heart valve appears to have disseminated from its primary focal point. Or the patient may have acquired a fungal infection during a previous surgical procedure. Correct identification and immediate recovery of the fungus from the direct smear or culture is essential. In image A, the cells are strongly stained as Gram positive. The structures seen clearly demonstrate budding forms that display elongation, mimicking pseudohyphae. The direct smear results were reported to the clinician as "yeast and pseudohyphae." A few days later the culture grew mold, identified as *Scedosporium apiospermum*. Did we make any mistake in interpreting the direct smear results? No, we did not. We have experienced that many times true mycelium producing filamentous form appear as yeast and pseudohyphae in direct smear. Upon isolating mold from the culture, the direct smear was re-checked. The cells attached to the tip of fungal hyphae did not look distinctly like true hyphae. The cells are elongated, some oval and bulgy in the central area, and they strongly stained Gram positive.

In image B, once again the structures are strongly Gram positive and shooting from a round cell looking like yeast. The structures produced appear as pseudohyphae. Therefore the results were reported to the clinician as "yeast with pseudohyphae." Two days later, the culture grew blue-green mold and was identified as *Aspergillus fumigatus*. Did we make a mistake in reporting pseudohyphae results from direct smear specimen? Yes, we did. The smear results were sent prematurely without noticing few clues that are clearly shown in the image. Looking closely at the hyphae it was noticed that the septation of the filaments is frequent in nature, and there is no pinched-off morphology seen at the segments. The walls of the hyphae are parallel, and the round cell that was thought to be a yeast is actually a "fungal conidium" that started to germinate and branched into septate hyphae.

The set of images in this quiz is interesting. Both structures stain strongly Gram positive. This reaction is often seen with yeast and pseudohyphae but not in true mycelium (or seldom so, as in this example). The reason has previously been explained in the beginning (Gram stain mechanism) and is that the cell walls of both images have allowed entry to the primary stain (CV) that was retained in the interior of the cells and staining as purple. However, the addition of iodine locked up CV within the cell behind the cell wall, making the CV-iodine molecules much larger. The application of decolorizer did not produce wider pores in the cell walls of the fungal hyphae as a result, and the decolorizer failed to remove the CV-iodine complex from the cell that gave the Gram reaction as purple (not truly a Gram positive reaction).

For **reporting** purposes image **A is okay as reported**. For image B, the direct smear results should have been amended and reported as: "**Septate hyphae seen.**"

20.3 Quiz 3

A BAL specimen from a lung-transplant patient (A) and a swab from scalp lesion (B) were submitted to microbiology for C&S and fungal investigation. Gram images taken at x1000 are posted below. Examine the structures and describe the statement you would issue to the physician (ward).

Image set Quiz 3

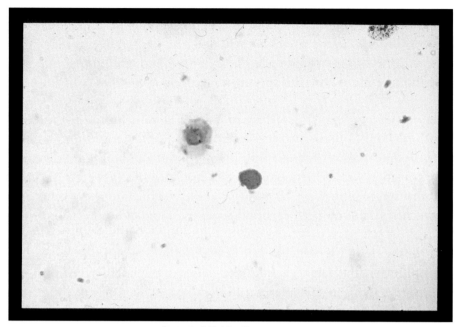

Quiz 3-A BAL: Gram x1000

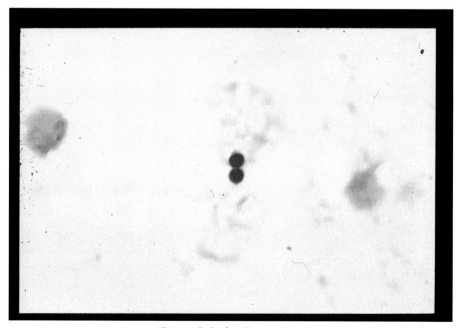

Quiz 3-B Scalp: Gram x1000

ANSWER TO QUIZ 3

Both images are similar in morphology. For image A, note the patient data (post-lung-transplant recipient). It means that the patient is at high risk of inhaling infectious agents (bacteria, viruses, fungi, and other microorganisms). Structures seen under the microscope must be thoroughly searched for clues in order to relay a meaningful interpretation to the clinician. The cells are irregularly round and non-budding, staining Gram negative. A closer look would reveal that these structures are not yeast but are conidia of fungus that are about to begin isotropic expansion to germinate. In these situations, when structural morphology does not fit into a specific type, describe structures as they appear under the microscope without naming a definitive fungal entity. However, in the comment field one may wish to describe all scenarios surrounding the structures seen. The correct interpretation would be: "Single, round, and non-budding cells mimicking fungus." The corresponding culture from this specimen recovered *Aspergillus fumigatus*.

In image B, the cells seen under the microscope are round and closely sitting together and were reported as budding yeast. This turned out to be an incorrect interpretation. The site of the specimen is a scalp lesion, indicating that the infectious process has taken place in a superficial site. No growth was obtained from the C&S. Fungal investigation was also requested for the specimen. Fungus grew well after a week as slow-growing white mold that was identified as *Trichophyton tonsurans*. The Gram smear has enough information to rule out yeast, which had already been reported from the Gram smear. A closer look would reveal a septum in between the two cells (a faint pinkish line), indicating the cells in compartments; i.e., septation. The reason it appeared as budding yeast is that part of the interior of both cells was stained strongly Gram positive, leaving the septum and the outer cell wall as unstained or stained light pink. The direct smear results were corrected and reported as: "Septate hyphae seen."

20.4 Quiz 4

A necrotic tissue from an abscess in the back of pharynx (A) and CSF (B) specimens were sent to microbiology for C&S and fungal investigation. Direct smears were stained by melanin stain (Fontana Masson) in the histopathology lab and read by the mycology technologist. Comment on the images taken at x1000 magnification.

Image set Quiz 4

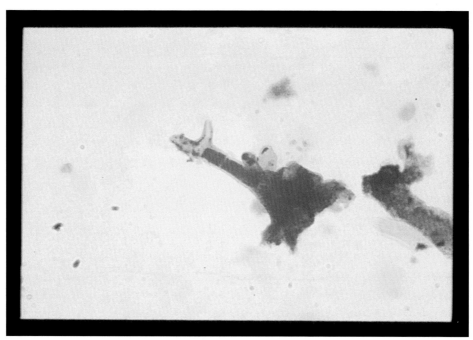

Quiz 4-A Oral Abscess: FM x1000

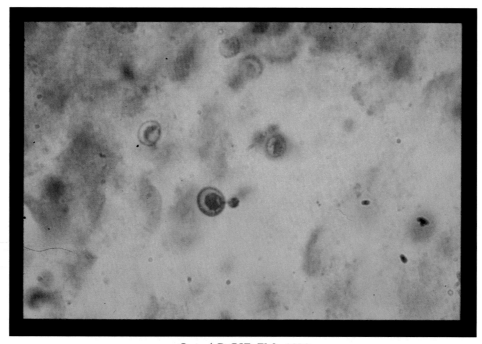

Quiz 4-B CSF: FM x1000

Answer to Quiz 4

In image A, the Gram smear comes from a necrotic tissue from the back of pharynx of an immunocompromised host. The Gram smear results were interpreted correctly as "Septate hyphae seen." However, the qualifying statement that was issued with the Gram smear result as "dark-walled fungal elements belonging to dematiaceous fungus" turned out to be incorrect. Many times phaeohyphomycotic agents were spotted from direct Gram smears or wet prep such as KOH, as were also seen in the above specimen during direct examination of the fungal elements. However, the darkness of the fungal elements seen required confirmation for melanin pigment in the cell wall by a special Fontana Masson (FM) stain. Within the next three days a yellow-green fungus hyaline in nature was isolated from the corresponding culture. No pigmented fungus was isolated. It was a total surprise, since the fungus recovered from the corresponding culture did not match the fungal elements observed in the direct Gram smear. Meanwhile, the special stained slide (FM) was ready to be examined, and no dark pigment on the fungal cell wall was found under the microscope. This led to a review of the Gram smear, and it was found that the interior of the fungal elements were on the darker side, but the delineated cell walls remained unstained; i.e., hyaline. Upon consulting the clinician about the gross appearance of the specimen, it came to be known that the specimen was darkly stained with dead tissue and masked with old and dead RBCs. The area was grossly necrotic. The fungus was deeply suspended within the colloidal and specimen debris, which may have been sucked in due to the decaying process, making the fungi darker. It is important for the clinician to know if the invading fungi are hyalohyphomycotic or phaeohyphomycotic in nature. A shift in selecting the appropriate antifungal agent may take place based on the additional information provided in reporting on fungal elements seen under the microscope.

In image B, the single cell stands on its own, stained by FM, which demonstrates the reaction positive for melanin pigment. The entity of this fungus is such that it is seldom necessary to demonstrate the presence of melanin pigment in this type of fungus, although it produces melanin pigment that contributes pathogenicity of this organism. However, this test was needed when other simple tests did not immediately give a positive result. The CSF specimen comes from an HIV-positive host; round cells were seen in the Gram-mimicking *Cryptococcus*. However, an India ink test did not show capsule, and mucicarmine stain was inconclusive. The only test remaining on hand to confirm the identity was melanin stain, which was done and demonstrated the dark cell wall of *Cryptococcus neoformans*. The FM stain displayed positive reaction for melanin.

20.5 Quiz 5

Examine the Gram smear images below (A) from BAL and (B) from a foot lesion.

Define the structures and state what fungus you would be expecting to recover from the corresponding culture.

Image set Quiz 5

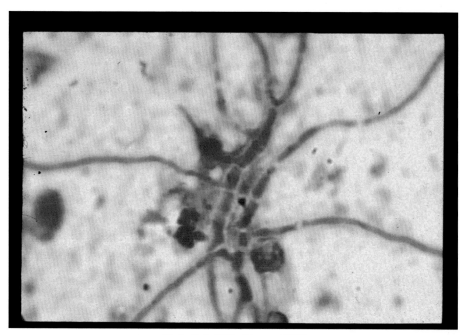

Quiz 5-A BAL: Gram x1000

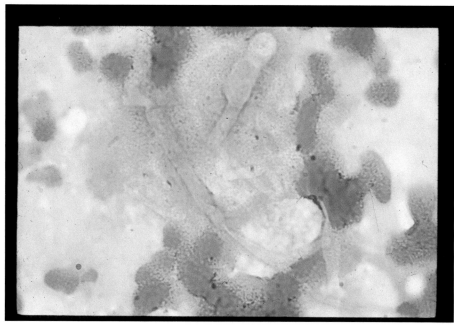

Quiz 5-B Foot Lesion: Gram x1000

Answer to Quiz 5

The Gram smear in A shows branching hyphae. Some hyphal structures display irregularly parallel walls, and the branching pattern demonstrates that it is dichotomous. This kind of branching pattern is often called "acute-angle branching." This tells us that the corresponding culture should recover a hyalohyphomycotic agent, most probably *Aspergillus*. The corresponding culture grew *Aspergillus fumigatus*.

In image B, the source of the specimen is a foot lesion. Fungal elements, when suspended deeply within the colloidal material of the specimen debris, often bar staining reagents from entering through the cell wall. As a result, the fungal elements seen under the microscope are not clearly visible, rendering the smear negative for fungus. However, a closer look would reveal the presence of fungal elements in segmented pattern. A filamentous form has also been displayed. Although the structures resemble pseudohyphae, they don't have individual budding forms and pinched-off morphology. For this reason, the structures were interpreted as "Septate hyphae." The corresponding culture media recovered *Fusarium* species.

20.6 QUIZ 6

Two Gram smear images below belong to two specimens: (A) tissue culture; (B) conjunctiva. Examine the images and describe what structures you would report.

Image set Quiz 6

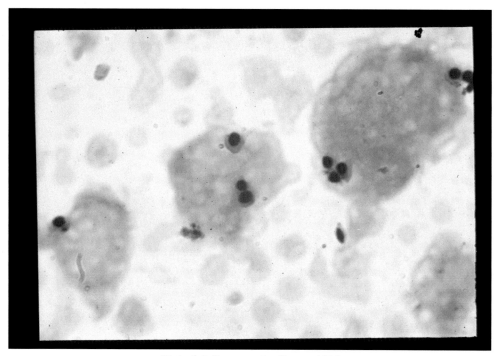

Quiz 6-A Cytogenetic: Gram x1000

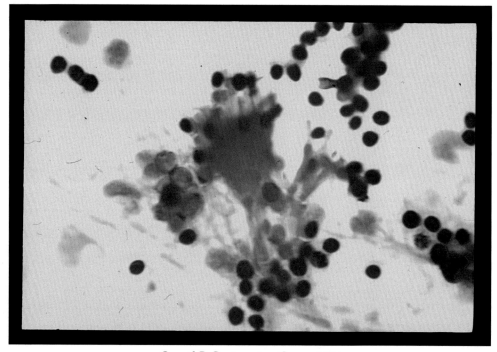

Quiz 6-B Conjunctiva: Gram x1000

ANSWER TO QUIZ 6

The two Gram smears made from A and B specimens are morphologically very similar. Both show small round cells mimicking yeast. The Gram stain results from both smears were reported as "yeast seen." However, the corresponding cultures from both specimens grew blue-green mold that were identified as *Aspergillus fumigatus* from image A and *Penicillium* species from image B. Upon isolating molds from culture media, both Gram smears were recovered and reviewed. A closer look allowed us to believe that these small round cells appearing in budding form are conidia. Some conidia were found sitting in close vicinity and appear as budding. The direct smear results were amended as "Fungal conidia seen."

20.7 QUIZ 7

Examine the following images taken from two different patients: (A) mediastinal biopsy and (B) subphrenic abscess. Record your findings and describe structures leading to the expected culture recovery.

Image set Quiz 7

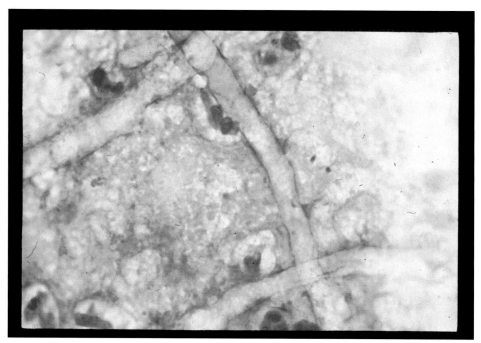

Quiz 7-A Mediastinal Biopsy: Gram x1000

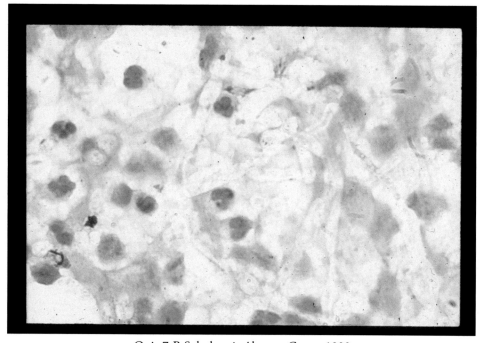

Quiz 7-B Subphrenic Abscess: Gram x1000

ANSWER TO QUIZ 7

Image A shows hollow unstained tracks of hyphae staining pinkish at the periphery. Initially Gram stain results reported as "NBS". What can you make from the Gram reaction of this image? Would you call the structures Gram negative? Let's find out what the true Gram reaction for this image actually is. Earlier you learned about fungal cell walls being thicker and tough. Fungal cell walls are also hydrophobic. As a result fungi would not allow water-soluble material to travel across the cell wall unless the cell wall of the fungi has previously been damaged. This image is an excellent example of fungi not taking up any basic Gram stain dyes (CV and safranin). As a result, the counterstain safranin sticks outside the boundary of the cell wall, making the fungal elements appear pinkish or Gram negative. However, these fungi actually are hollow structures of unstained hyphae. The cell walls are parallel and do not show any constrictions along their length. Septation is barely visible and tough to locate. There are no yeast cells seen in this image (single or budding). The structures were reported as "Septate hyphae seen" and the corresponding culture grew *Aspergillus fumigatus* within a few days. Although the structures are unstained, the counterstain sticking to the side of the filament provides a contrast to the microscopic preparation, making the detection of the fungal structures easier. The Gram stain reagents did not stain these structures; however, it indirectly helped detection of unstained fungal elements.

The structures seen in image B are even tougher and more difficult to interpret. The original Gram smear was reported as "no fungal elements seen." Upon isolating yeast from the corresponding culture media, the smear was reviewed. Upon spending much time under the microscope, some fungal elements that were initially invisible started to form a shape. Looking closely, fungal structures began to show round forms attached to one another. They also formed some filaments. It was very hard to determine septation. However, some cells showed budding and constrictions, indicating the morphology of the structures as yeast and pseudohyphae. The corresponding isolate form culture was identified as "*Candida albicans*." Both images in this set have the same explanation, that fungal elements, when masked by colloidal material of the tissue components, do not allow Gram reagents passage across the cell wall. As a result, fungal elements remain unstained with an invisible ghost-like pattern.

20.8 QUIZ 8

Image (A) below is a Gram stain made from mandible lymph node, and image (B) comes from a BAL specimen from an immunocompromised host. What would you tell the clinician about Gram smear results pending culture confirmation?

Image set Quiz 8

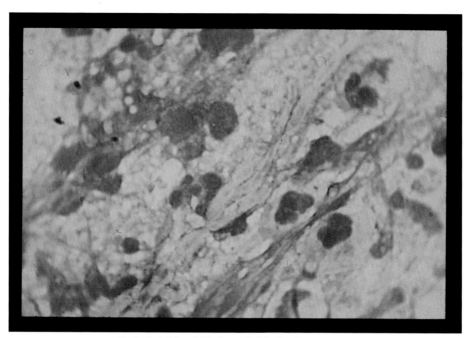

Quiz 8-A Mandible Lymph Node: Gram x1000

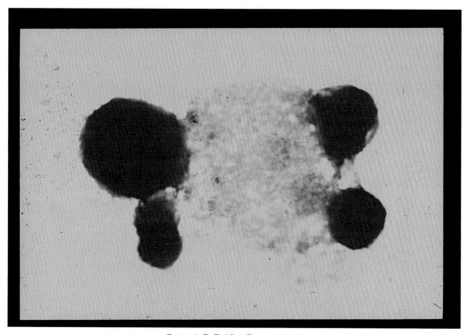

Quiz 8-B BAL: Gram x1000

Answer to Quiz 8

Image A from this set is similar to the one in Quiz 7 in appearance as unstained hyphal fragments with inadequate contrast. The specimen contained massive amounts of fungal hyphae that appeared as septate and pseudohyphae without clarity. The smear results were defined as they appeared under the microscope. However, no growth was obtained after weeks of incubation. The patient was immunocompromised and also on antifungal therapy, which may have retarded the viability of the fungal elements and suppressed its growth in culture media. The specimen was sent for molecular studies, where it was identified as *Candida albicans*.

In image B, we see some host cellular response as well as larger cells in the central area surrounded by host cells (PMNs?) showing faintly staining pale pink color. The central cell is a macrophage or similar in the phagocytic category. The central cell is surrounded by polymorph nuclear leukocyte-like cells, indicating that phagocytosis against the invading organism is in action. This fungus was easily seen fluorescing under the UV microscope during examining a CW-stained smear. Small oval budding cells were seen under the UV microscope, but no cellular material was seen, since CW does not stain host cellular material. The initial Gram smear results were reported as "pus cells seen; no bacteria seen." Upon seeing budding yeast cells in the CW smear, the Gram smear was reviewed and structures found, as shown in image B. The patient was suffering from pneumonia and was HIV positive. This helped us to assume the presence of intracellular forms of *Histoplasma* in the Gram smear. Note the clusters of small oval cells unstained and dull-looking packed inside the large pus cell in the center. A white mold was isolated later on from the corresponding culture, identified as *Histoplasma capsulatum*, and confirmed by DNA studies at the reference lab. The Gram smear results were amended as "Small oval budding yeast cells, intracellular, suggestive of *Histoplasma*."

20.9 Quiz 9

A sputum Gram x1000 (A) and an abscess aspirate KOH x400 (B) display fungal elements. Identify the entity of these fungi and report direct smear results to the clinician.

Image set Quiz 9

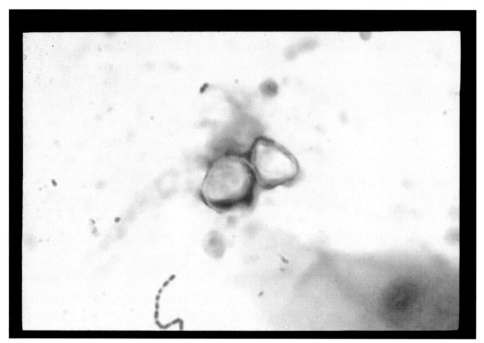

Quiz 9-A Sputum: Gram x1000

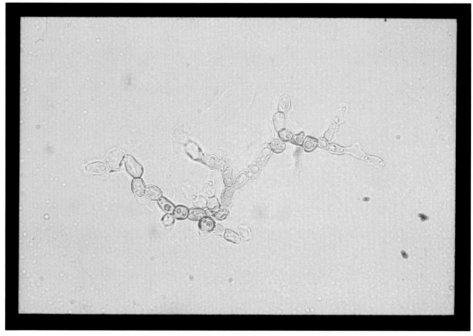

Quiz 9-B Abscess Aspirate (foot cyst): KOH x400

ANSWER TO QUIZ 9

Image A from sputum shows two round cells "hugging" each other. There is no other detail available, except that the cell wall is thick, the cell size appears above 10 μm, and the shape is more-or-less round. This is not a classic picture of *Blastomyces*. However, there are sufficient minor details available on hand to suspect that the structures are *Blastomyces*. At this point, a simple wet prep or KOH would be most useful to demonstrate round cells having double wall with broad base, although the image does demonstrate broad base. However, in this case the cytoplasmic stream going from the mother cell to the daughter cell via the broad base is missing. Without such important difference, it would be very hard to confirm the structures as *Blastomyces* or two immature spherules of *Coccidioides* sitting side by side and overlapping a part of the cell walls of both cells. The culture grew a white mold and was identified as *Blastomyces dermatitidis* after 11 days. For reporting purposes the structures may be interpreted as "large, round, thick-walled budding yeast with broad base suggestive of *Blastomyces*." In this case, the travel history of the patient would provide extra information useful in making a decision to rule out *Coccidioides*, since the patient did not travel to any endemic area of *Coccidioides*.

Image B of this set is taken from a KOH prep after the structure seen in the CW and Gram smear appeared to be different in terms of pigmentation in the fungal cell wall. The CW stain does not display the pigmented cell wall of dematiaceous fungi. Therefore, it was decided to examine the structure under the bright field to rule out dematiaceous fungus. In this image we would find structures slightly pigmented along with the morphology of the cells seen appearing to be in chains but not truly in budding form. Structures demonstrate a caterpillar type of constrictions at segments known as "toruloid" morphology, making hyphal fragments appear swollen. Such structures are most often seen in dark fungi. The results were reported as "septate hyphae dark-pigmented belonging to dematiaceous group seen." Culture grew *Exophiala jeanselmei* after 10 days of incubation.

20.10 Quiz 10

A positive blood-culture bottle was processed and an image was taken from the Gram smear (A). Another Gram smear was prepared from the BAL (B) specimen. What do you see in these Gram smears?

Image set Quiz 10

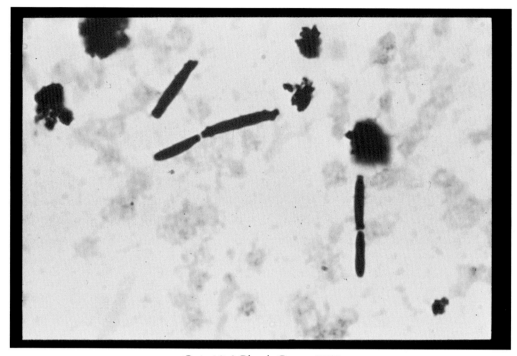
Quiz 10-A Blood: Gram x1000

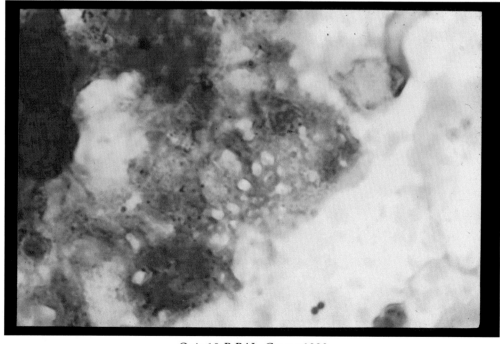
Quiz 10-B BAL: Gram x1000

ANSWER TO QUIZ 10

Image A from blood culture shows a long hyphal fragment, somewhat constricted at the joining point, and a space is left in between the two elongated cells. There were no budding yeasts seen except the short hyphal fragments. These structures should be considered as yeast-like morphology and not the septate hyphae of true molds. The space may indicate the presence of a septum; however, it will be confusing to the clinician if such structures are called septate hyphae. Whenever septate hyphae report is seen by the clinician, the first thing that comes to mind is *Aspergillus*, followed by other hyphomyctes. The structures seen in this Gram are not true fungi but are yeast-like fungi. The structures seen in Gram smear were reported as "septate and pseudohyphae seen." The culture grew yeast-like fungus within the next few days and identified as *Trichosporon beigelii*, which is a yeast-like fungus and produces septate hyphae as well as pseudohyphae and blastospores.

Image B from the BAL shows interesting structures. The Gram image here depicts the morphology of intracellular structures. However, *Histoplasma* is ruled out, although the patient was HIV positive. There are several round bodies; most are unstained, and others have a pink dot located centrally. If you ever happen to see structures resembling the morphology of this image, do not give up on it. Make arrangements to get the specimen tested for PCP using either GMS, CW, or IFA. The structure shown in this image is "foamy exudate" that is produced by the interstitial fluid in the lung in patients suffering from PCP. The PC cysts are suspended within the foamy exudate, giving an appearance of a "honeycomb." *Pneumocystis jirovecii (carinii)* was reported from this specimen after confirming PC cysts by CW (Fungi-Fluor by Polysciences).

20.11 QUIZ 11

A BAL specimen from a post-lung-transplant patient was received by the lab; C&S and fungal culture were asked for. Examine the Gram smear (Image A), fungal smear (Image B), followed by the Gram smear (Image C) and comment. The clinician is suspecting *Aspergillus* in the specimen.

Image set Quiz 11

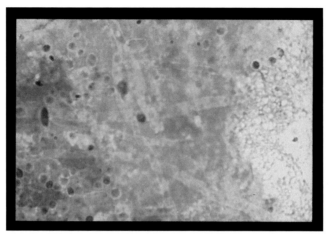

Quiz 11-A BAL: Gram x1000

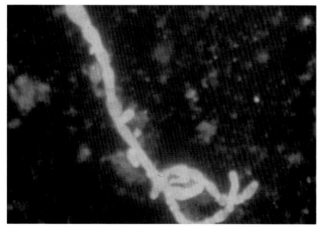

Quiz 11-B BAL: CW x400

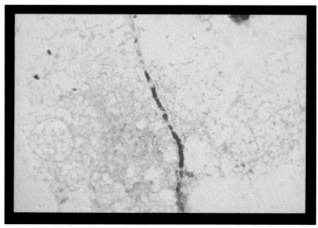

Quiz 11-C BAL: Gram x1000

ANSWER TO QUIZ 11

When examining direct smear under the microscope, the smear reader must develop a strategy not only to report microorganisms to the clinician but also to provide the interpretation so that the receiver on the other side gets a clear message. Upon receiving the lab results the clinician puts the patient on appropriate therapy. If for some reasons the interpretation of the microscopic objects seen is inaccurate, a corrective action is necessary to amend the situation. In this example, the objects seen under the microscope were not clear cut. The yeast and pseudohyphae are usually picked up by Gram stain; however, in the above case, it was not observed until after the Gram smear was reviewed (Image C). The fungal smear (CW) showed excellent structures of yeast and pseudohyphae morphology. Culture grew yeast that was identified as *Candida albicans*. No *Aspergillus* was isolated nor was suspected from the direct smear examination.

The laboratory technologists are not responsible for determining the clinical significance of the microorganisms; however, the laboratory staff provides valuable information capable of determining the legitimacy of the microscopic objects seen under the microscope. Therefore, the lab staff must be able to determine the fungal elements seen as real and coming from the specimen. This can be easily done when the objects seen under the microscope match the isolate recovered from the corresponding culture media.

In the above example, the clinician suspected aspergillosis. However, no mold was isolated except *Candida albicans*, which matched with the direct smear results.

20.12 Quiz 12

A BAL specimen from a post-lung-transplant patient was received to be analyzed for C&S and fungus. Gram results were reported as "pus cells seen; NBS." Examine the fungal smear (Image A) followed by the original Gram smear (Image B) and a GMS smear (Image C). The clinician was suspecting aspergillosis. The culture grew mold after a few days, and a wet prep using LPAB was made and examined under the microscope (Image D). Discus the direct microscopy and the culture recovery.

Image set Quiz 12

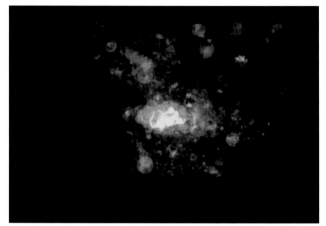

Quiz 12-A BAL CW x400

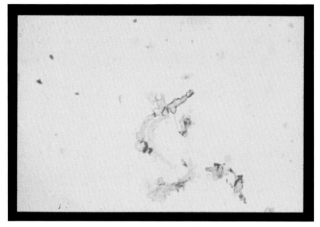

Quiz 12-B BAL Gram x1000

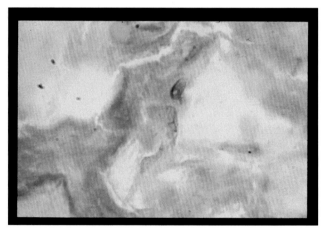

Quiz 12-C BAL GMS x1000

Quiz 12-D BAL LPAB x400

ANSWER TO QUIZ 12

Fungi seen in direct KOH or CW smears prepared from the clinical specimens are usually present in the Gram-stained smear. The ability of the smear reader to detect fungi in the direct Gram smear would have a positive impact on the patient care. When fungi are present in the Gram smear but not picked up by the smear reader, review of the Gram smear is necessary (refer to section "reasons for missing FE in Gram"). In this case, two things became apparent: 1) Fungi were seen in the Gram upon review; and 2) the fungal elements seen did not appear to be *Aspergillus*. Organ transplant (especially lung) recipient patients usually acquire aspergillosis, although any fungal species may cause fungal infection in the lungs. Image A shows short filamentous forms in CW x400 that do not display dichotomous branching, a clue that is most frequently used to suspect the involvement of *Aspergillus*.

The Gram smear in image B showed septate hyphae; however, the structures lacked clarity in order to suspect fungal elements as *Aspergillus*. A few days later the fungus appeared on culture media and was identified as *Scopulariopsis brumptii*. *Scopulariopsis* is usually a nonpathogenic fungus; however, this fungus has been involved in causing onychomycosis. Organ-transplant recipient patients are kept on immunosuppressive therapy; therefore, these patients become immunocompromised on purpose, and as a result they are vulnerable to acquiring a variety of infections, most importantly aspergillosis. This case signifies that a normal nonpathogenic or less-virulent fungus is capable of causing fungal infection in patients whose immune state has been compromised either by disease, medication (other factors), or on purpose.

20.13 Quiz 13

A wound swab received from a 72-year-old female patient was submitted for C&S. The Gram smear results reported to the clinician as "few pus cells, few bacteria, and few yeast" (Image A). No fungal culture was requested on the specimen. The C&S culture grew heavy commensal flora and a light growth of mold identified by LPAB (Image B). What would be your next step before reporting fungal results to the physician?

Image set Quiz 13

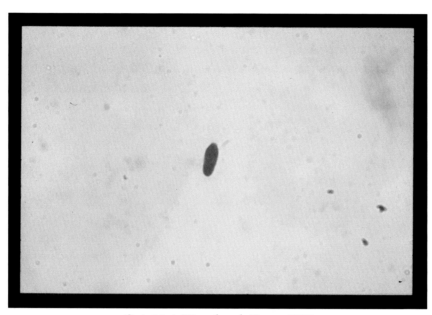

Quiz 13-A Wound swab Gram x1000

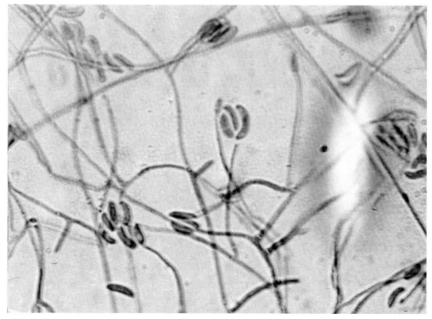

Quiz 13-B Wound swab LPAB x400

ANSWER TO QUIZ 13

The patient's immune status may have been temporarily and/or locally impaired. As a result, the fungus was able to dig into the tissue, causing infection. The overwhelming majority of fungi are usually nonpathogenic or they are all opportunistic, since fungi do not specifically depend on human tissue for nutritive requirements (fungi absorb food from decaying organic matter). Therefore, any fungus capable of causing infection must have passed three barriers (the temperature, immune status, and complex organic tissue material) before invasion. In the above case, the fungal elements were seen in the direct Gram smear but were interpreted as yeast due to the shape they produced. Incidentally, the clinician did not request fungal investigation on the patient's specimen and asked for C&S only. As a result, yeast reported from an initial Gram-stained smear remained unchecked until after the mold grew from culture and was referred to mycology for identification. Upon receiving the mold for identification, the mycology technologist looked at the direct smear results, which did not match the culture recovery of the fungus. The Gram smear was reviewed, and it was found that the yeast-like cells seen and reported from the Gram stain were not yeast but conidia (macroconidia) of a fungus. The fungus isolated was identified as *Fusarium* species by examining a wet prep under the microscope, using LPAB. Whenever there is discrepancy between the direct smear results and the culture, always review the direct smear. Many times such errors are easily rectified.

20.14 QUIZ 14

A 41-year-old female cancer patient developed lesions in the scalp. The physician suspected opportunistic fungus, and a biopsy specimen from the scalp was sent to microbiology for C&S and fungal culture. The Gram smear results were reported as "no pus cells seen & NBS." C&S was reported as "no growth obtained." Examine the fungal smear (CW) prepared from the biopsy specimen (Image A) and also the KOH (Image B). Would you be able to identify fungus seen in direct smear at least to genus level before looking at the culture morphology (Image C) and LPAB prep (Image D)?

Image set Quiz 14

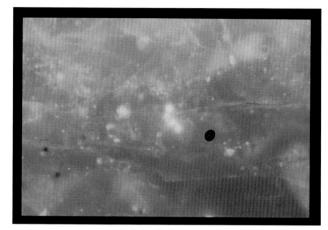

Quiz 14-A Scalp biopsy CW x400

Quiz 14-B Scalp biopsy KOH x400

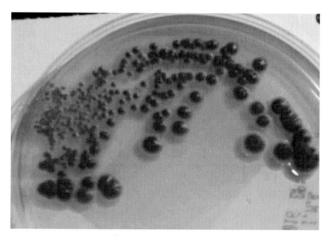

Quiz 14-C Scalp mold growing on IMA

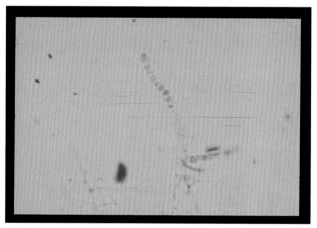

Quiz 14-D Scalp biopsy LPAB x400

ANSWER TO QUIZ 14

The patient is suffering from cancer and is on cortisone (i.e., is an immunocompromised host). The lesions developed in the scalp area prompted microbiological analysis with suspicion of opportunistic fungal infection. The direct Gram smear was negative; however, the fungal smear in the mycology lab showed very few fungal elements that appeared to be chlamydoconidia of *Trichophyton verrucosum* or *Trichophyton violaceum*. The smear was re-examined thoroughly, and the few fungal elements observed were only seen in the periphery (outer part) of the specimen, leaving the entire deep side of the tissue unaffected. This indicated that the nature of fungus must be restricted from entering deeper areas of the tissue. This prompted us to suspect the presence of a dermatophytic agent in the specimen. The specimen was also viewed in KOH prep, which showed structures exactly like the CW prep but with more clarity.

One must keep in mind that the CW stain binds the chitin and cellulose in the cell surface (the cell wall) and optimizes visual clarity to make structures brighter under the fluorescent light; however, CW does not stain the cellular components to enable viewing internal details of the fungal cell. The KOH serves as an excellent alternative to view the detail of the fungal cell since light passes through the cell forming image(s).

The Gram smear was also reviewed, and no fungal elements were observed. A week later the culture grew small granular colonies more or less in a circular fashion and having a fringed periphery. A quick LPAB prep made and examined under the microscope showed structures exactly like the direct smears (CW & KOH). Upon continued incubation, the mold colonies produced reddish and purplish pigment consistent with *Trichophyton violaceum*. A preliminary report was issued pending the final word from the reference lab and confirmed the isolate as *Trichophyton violaceum*. The culture did not grow at 37⁰C, ruling out *T. verrucosum*, which usually remains non-pigmented. The key area of discussion is to know that fungi usually do not produce conidia in human tissue unless the environment in certain body sites become equivalent to the natural state for fungi to reproduce by conidiation. It is for this reasons that many fungi do not go beyond the superficial site and are restricted from going deeper into the human body, since they are not capable of surviving the host defense mechanism. This case represents the natural environment for the fungus provided by the superficial site (scalp), allowing the fungus to produce conidia that classically fit into the *Trichophyton verrucosum* and *Trichophyton violaceum* groups.

20.15 Quiz 15

A BAL specimen collected from a post-lung-transplant-recipient female patient was sent to the microbiology laboratory. The Gram smear results were reported as "pus cells and commensal flora and large yeast cells seen." However, "no growth" results were reported from C&S. The corresponding fungal smear in mycology showed thin-walled round cells. No growth was obtained after four weeks on incubation. Examine the Gram smear image (A), CW image (B), KOH image (C), and GMS image (D) and comment on the situation and provide the next step to resolve the matter.

Image set Quiz 15

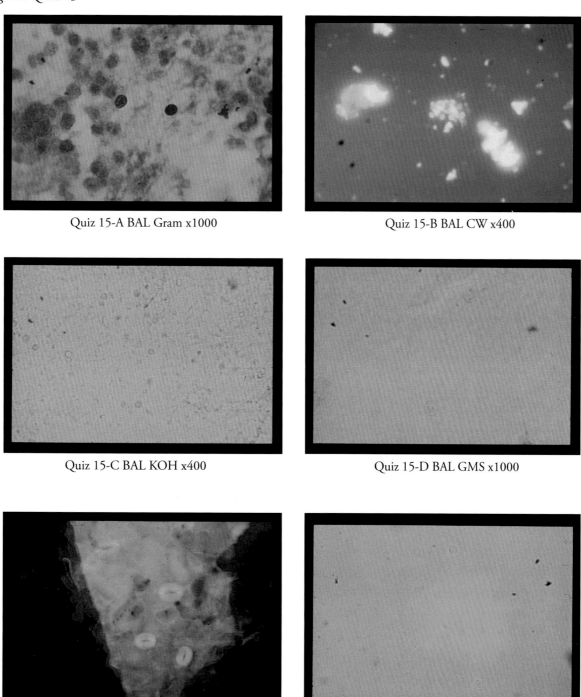

Quiz 15-A BAL Gram x1000

Quiz 15-B BAL CW x400

Quiz 15-C BAL KOH x400

Quiz 15-D BAL GMS x1000

Quiz 15-E BAL CW x400 (plant cells – respiratory cells known as "stomata").

Quiz 15-F BAL KOH x400 (possibly other plant cells)

ANSWER TO QUIZ 15

According to the physician, this patient was immunocompromised and very sick. They desperately needed a clue to treat the patient; however, nothing conclusive was available from the microbiology or pathology departments. The morphology in the Gram smear results reported as large "yeast cells seen" was not enough to conclude as *Candida, Cryptococcus,* or *Blastomyces,* since all three could produce large yeast cells. The CW smear done in mycology clearly showed thin-walled, small to medium-sized round cells but did not specifically show any other evidence to categorize such yeast cells into the three different groups of fungi as mentioned above.

These are the situations when the lab finds itself stuck and incapable of producing useful information to the clinician. There are a few important things to consider. The Gram stain showed cells resembling yeast; however, a closer look at the cells would cast some doubt since the cells do not strongly appear purple as they usually stain by Gram's method. The next thing to notice is the CW (image not available), which clearly showed small to medium-sized round cell, including few budding forms. The KOH showed similar structures to CW, except that the prep was unstained. At this point, mycology direct smear results were sent out as "thin-walled yeast cells seen" just to allow the clinician not to confuse with Gram smear results that were reported out as large yeast cells seen.

When particular clues are missing in direct smear morphology of the cells, no qualifying statement such as identification of the organism should be included. The clinician came back for further information; however, nothing else was available to be added, due to the lack of other details. At this point, a fresh smear was prepared and stained by GMS. The GMS smear turned out to be more confusing than the primary simple stains (Gram, CW, and KOH), since nothing looking like fungus was seen in the GMS smear. Upon further searching, only faint, round and oval unstained cells were observed. The cells seen appeared as tissue structures or plant cells, but this could not be confirmed due to the insufficient quantity of the specimen.

The pathology lab report was also negative. The pathology department was requested to forward any remaining specimen or send it directly to the reference lab where a PCR test may be performed to rule out fungus and or *Candida/Cryptococcus.* No further follow up was received, and nothing conclusive could be said about the structures seen under the microscope. If the structures seen are really plant cells, where would they have come from? Could the answer to this question be food (stomata & other plant cells – E & F: Image set Quiz 15) that may have been inhaled?

The images E & F do not belong to the patient's specimen images. These images were taken from a different patient's BAL specimen who swollen and inhaled green vegetable during meal.

20.16 QUIZ 16

A BAL specimen was collected from a 35-year-old female double-lung-transplant recipient and sent to microbiology for C&S and fungal culture. The Gram smear results were reported as "pus cells seen, NBS," and the culture results were reported as "yeast not *Candida* or *Cryptococcus* isolated." The corresponding fungal smear showed yeast-like cells (Image A) and were reported as "fungal elements seen" pending culture recovery of the isolate. Comment on the fungal smear results and also Gram smear (Image B). What would you have reported from the Gram smear if these cells were picked up by the smear reader the first time?

Image set Quiz 16

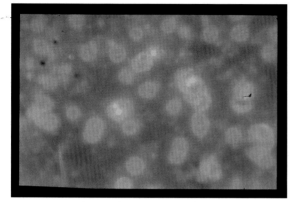

Quiz 16-A BAL CW x400

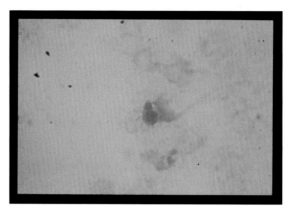

Quiz 16-B BAL Gram x1000

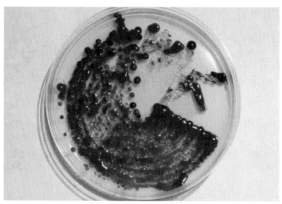

Quiz 16-C BAL mould on IMA

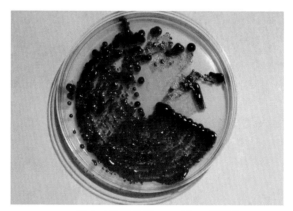

Quiz 16-D BAL mould on EBM.

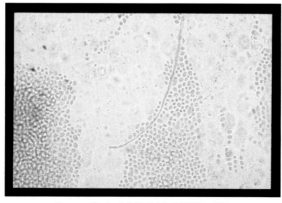

Image Quiz 16-E BAL LPAB x400

Answer to Quiz 16

This case illustrates the insight based on experience. Before we discuss insight and experience, we must reiterate that constant observation is essential over time. It is for this reason that the cells seen in the fungal smear (CW) appeared budding and yeast-like; however, the term "fungal elements" was used in reporting instead of budding yeast. The structures seen in the fungal smear did not appear to be true budding forms.

C&S culture plates (blood agar and chocolate agar) were received for yeast identification and incubated at 28°C. All media grew shiny mucoid and dark colonies (Image C and D), and *Exophiala* was suspected. A quick wet prep using LPAB[20] was prepared and examined under the microscope (Image E). *Exophiala dermatitidis* was confirmed by DNA sequence carried out at a reference center. Upon review of the original Gram smear, structures seen appeared to have a septum. No budding yeast was observed in the Gram smear. The patient was on antifungal therapy (voriconazole), and the clinician was suspecting aspergillosis. Dark fungi (dematiaceous group) produce melanin pigment that contributes to virulence. This information is often useful early on in order to select an appropriate antifungal agent. No *Aspergillus* was isolated from this specimen.

20.17 QUIZ 17

A 71-year-old female patient was admitted to the hospital suspected of having Cryptococcal infection. Blood was sent in BacT/Alert bottles (automated system) for culture and sensitivity testing. The blood-culture bottle became positive within 48 hours, and a Gram stain was prepared and examined, which revealed the presence of round and oval yeast cells (Image A). The physician questioned the yeast cells that were not clearly fitting in *Cryptococcus* morphology. A wet prep from the blood culture was made and examined (Image B). The positive blood culture was processed, and mucoid colonies mimicking *Cryptococcus* appeared on blood agar and chocolate agar (Image C). The culture was processed for yeast identification. IMA and EBM showed mucoid light grayish colonies (Image D). Cornmeal morphology showed *Candida krusei* like structures (Image E). API 20C AUX identified the isolate as *Rhodotorula* with 99.9% confidence. Would you trust API 20C AUX ID or cornmeal morphology? What would you tell the clinician in your preliminary statement?

Image set Quiz 17

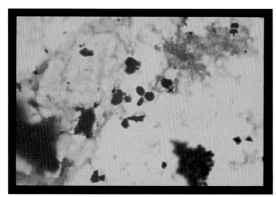

Quiz 17-A1 Blood Culture: Gram1 x1000

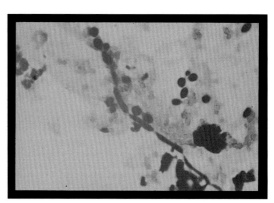

Quiz 17-A2 Blood Culture: Gram2 x1000 re-done at day 7

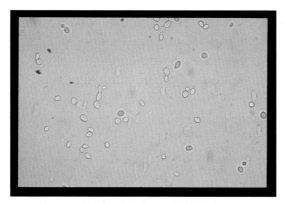

Quiz 17-B Blood Culture: KOH x400

Quiz 17-C Blood Culture: growing shiny moist colonies on Chocolate agar

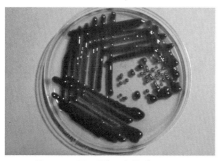

Quiz 17-D1 Blood Culture: growing brown, shiny & moist colonies on IMA

Quiz 17-D2 Blood Culture: growing brown shiny & moist colonies on EBM

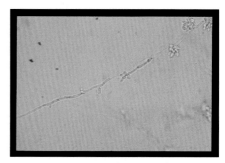

Quiz 17-E Blood Culture: Cornmeal Agar x400

ANSWER TO QUIZ 17

This case is interesting in many ways. The patient was immunocompromised, and the clinician suspected Cryptococcal infection. The blood-culture bottle turned positive and was processed on culture media including Gram stain. A Gram stain prepared from the positive blood culture showed round and oval-shaped cells in single, pairs, and short chains. The culture grew moist yeast that was forwarded to mycology for further testing. The clinician questioned the yeast that did not resemble *Cryptococcus*. A quick wet prep using KOH was examined under the microscope, and round and oval cells in pairs and short chains were observed. The clinician was notified about the yeast cells not exclusively fitting the morphology of *Cryptococcus*; however, one unique characteristic was observed, that the cell walls of all cells seen were "darker." This is the key characteristic of *Cryptococcus neoformans* and is usually observed in wet prep such as KOH.

Initially the culture grew pinpoint colonies on blood and chocolate agar plates. A mucoid shiny texture mimicking *Cryptococcus* was noticed. A germ tube that was set up turned negative. Yeast identification was set up using "cornmeal agar, urea, SAB, IMA, EBM[12], and API 20C AUX" and incubated at 28⁰C. All culture media and tests were read after a 48-hour incubation. The urea test was positive, indicating the possibility of *Cryptococcus*. However, the cornmeal agar morphology showed structures resembling *Candida krusei*. API 20C AUX identified the yeast as *Rhodotorula* with >99% confidence value. It is also urea positive; once again cornmeal defeats the identification achieved by API 20C AUX.

IMA and EBM grew mucoid watery colonies turning slight grayish pigment that later became dark brown with age. EBM is a medium containing esculin for the production of brown colonies by the enzyme phenol oxidase produced inclusively by *Cryptococcus neoformans*. IMA does not contain esculin but gave darker pigment exactly the same as it did on EBM. Therefore, the test to produce brown colonies by the presence of phenol oxidase enzyme failed for *Cryptococcus neoformans*.

It was not hard to suspect that the mucoid yeast belonged in the dematiaceous group due to the fact that it produced brown pigment in medium devoid of esculin such as blood agar, SAB, and IMA. The cornmeal morphology was re-examined, and it was found that the hyphal walls were on the darker side and pinched-off morphology was giving rise to annellides bearing single-celled and rare bisected conidia. At this point, it was clear that the isolate is a mucoid dark yeast. The preliminary report was issued *Exophiala* isolated and confirmed by the reference lab as *Exophiala dermatitidis*.

20.18 QUIZ 18

A thigh aspirate from a 50-year-old male leukemic patient was received by the microbiology lab for C&S and fungal analysis. The patient was on voriconazole. The Gram smear results were reported as "many pus cells seen, NBS" (Image A). C&S results were reported as "no growth obtained." The corresponding fungal smear in mycology showed fungal elements of pseudohyphae morphology (Image B). The original Gram was reviewed, and no fungal elements were observed. The original Gram smear was overstained by CW/KOH, and fungal elements were observed (Images C and D). However, after four weeks no fungus was isolated from culture media. What would you tell the clinician about fungal elements seen in direct smear that failed to grow in culture media?

Image set Quiz 18

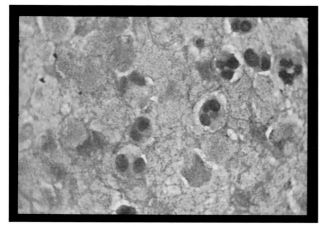

Quiz 18-A Thigh-aspirate Gram x1000

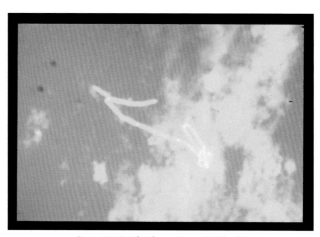

Quiz 18-B Thigh-aspirate CW x400.

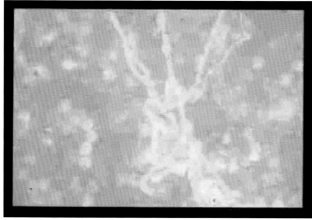

Quiz 18-C Thigh-aspirate Gram over-stained by CW-KOH x400

Quiz 18-D Thigh-aspirate KOH x400.

ANSWER TO QUIZ 18

Two main points were noticed in this case: the patient's depleted immune response and prophylaxis use of the antifungal therapy (voriconazole). The failure to detect fungal elements is a manual error due to one of the many reasons described in another section. Fungal elements seen in the direct fungal smear and also upon overstaining the original Gram smear using the KOH/CW procedure indicate that the fungal elements were detectable and identifiable. Yeast and pseudohyphae are easily detected by any microbiological-staining procedure. However, the organism failing to grow in culture media is a little odd.

No growth obtained after a full four weeks indicates that the organisms may have lost viability due to the use of an antifungal agent prior to collecting the specimen for culture. At least, the direct smear indicates that the fungal elements seen in the direct smear belong to a yeast-like fungus and not a mold. The lack of growth from the culture also confirms that the treatment is working and it may not be necessary to change the course of the patient's therapy.

20.19 Quiz 19

An abscess aspirate specimen collected from a sixty-six-year-old male sarcoma patient was submitted to the microbiology lab for C&S only. No fungal investigation on the specimen was requested. The Gram smear results were reported as 3+ pus cells and NBS (Image A). After forty-eight hours, one colony of yeast (Image B) was reported from bacteriology culture and referred to the mycology section of the microbiology department for identification. Examine the images and prepare an interpretative statement about the direct Gram smear results as well as commenting on the corresponding culture.

Image set Quiz 19

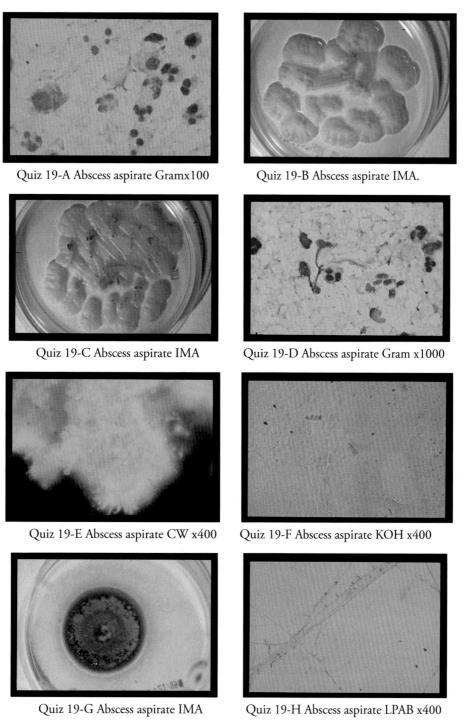

Quiz 19-A Abscess aspirate Gramx100

Quiz 19-B Abscess aspirate IMA.

Quiz 19-C Abscess aspirate IMA

Quiz 19-D Abscess aspirate Gram x1000

Quiz 19-E Abscess aspirate CW x400

Quiz 19-F Abscess aspirate KOH x400

Quiz 19-G Abscess aspirate IMA

Quiz 19-H Abscess aspirate LPAB x400

ANSWER TO QUIZ 19

In mycology lab, the culture morphology of the growth appeared as dull, pale, dry, and yeast-like. The battery of tests, such as cornmeal, urea, and API 20C AUX, were set up to identify the yeast to species level. Upon further incubation, the culture started to grow as glabrous, pale, dull, dry, raised colony that was subcultured on IMA for purity (Image B). After forty-eight hours, the cornmeal agar plate was examined. It showed morphology like filamentous fungus totally unrelated to yeast or yeast-like fungus. Upon extending incubation, a change in the colonial morphology was noticed (Image C) at day ten. The colonial morphology remained glabrous; however, the dark patches were noticed in the central area. This alerted mycology to review the Gram smear in search of fungal elements other than yeast. Upon carefully observing the Gram smear, a specific area suspicious for filamentous fungus (Image D) was detected. The low number of organisms seen did not appear obvious, and no pigmentation in the cell wall was observed. The culture that displayed spots of dark gray areas prompted to overstain the original Gram stained-smear using CW/KOH procedure. Under the microscope, the Gram smear displayed few fungal elements. The CW (Image E) stained-smear revealed structures not characteristically fitting in any particular group (yeast, pseudohyphae, or true mycelium). The author's experience suggests that many times, dematiaceous fungi behave differently in direct smear, making it difficult to categorize them. Although the Gram stain did not show pigment in the cell wall, the KOH (Image F) prep produced enough contrast to make the pigmented cell walls of fungal elements stand out in greater clarity. This was confirmed upon turning the mold (Image G) on IMA completely dark after two weeks, and lactophenol wet prep (Image H) identified the fungus as *Phaeoacremonium* species. *Phaeoacremonium* is structurally very similar to *Acremonium* and *Phialophora parasiticum*. The *Acremonium* genus belongs to the hyaline group and does not have a pigmented cell wall. *Phaeoacremonium* belongs to dematiaceous group causing phaeohyphomycosis.

20.20 QUIZ 20

Examine images (A, B &C) in the Gram, KOH & CW processed in mycology from a histology waxed frozen section of supraglottic tissue. Can you confirm the organism suggested by the histopathology team as *Histoplasma* or *Pneumocystis*? Noting that the background of the images is heavily masked by the wax that was not completely removed by the over staining procedure.

Image set Quiz 20

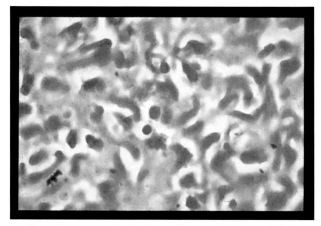

Quiz 20-A supraglottic tissue Gram x1000 waxed-slide

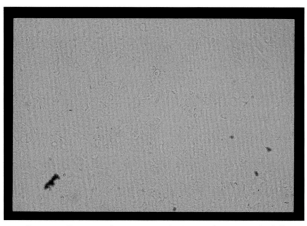

Quiz 20-B supraglottic tissue KOH x 400 waxed-slide

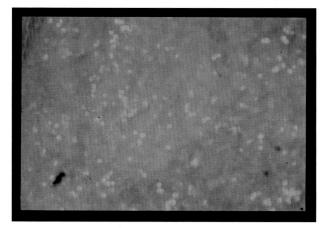

Quiz 20-C supraglottic tissue CW x 400 waxed-slide

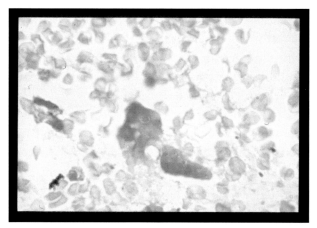

Quiz 20-D supraglottic tissue Gram x1000 waxed-slide.

ANSWER TO QUIZ 20

A fifty-year-old immunocompromised (HIV-positive) female patient was admitted to the hospital with chronic inflammation in the pharynx area. A tissue was collected from the supraglottic region and submitted to the histology department. The GMS stain done on the tissue revealed some suspicious cells appearing to be *Histoplasma* and/or *Pneumocystis*.

No specimen was sent to microbiology for C&S or fungal analysis. The clinician wanted the organisms seen in the smear be confirmed by mycology and requested the pathology lab to send their unstained slides to be processed by a mycology expert. The unstained smears were overstained using CW/KOH procedure. The smears were examined under the microscope, and the results were predictable—the Gram smear (Image B) did not show any organisms. However, the CW (Image B) and KOH (Image C) smears showed numerous organisms, the majority of them small and in oval morphology. Clear budding forms were seen in few cells. Due to the remnant oily material still interfering with the clarity of the organisms, it was assumed that the structure seen was *Histoplasma* and not *Pneumocystis*, since the histology results indicated the saucer-shaped cells most likely to be *Pneumocystis*.

The Gram smear was re-examined once again and carefully searched for clues. *Histoplasma* stains very poorly by Gram reagents; therefore, not finding enough organisms in the Gram smear alerts the smear reader about *Histoplasma* that is either poorly stained or completely hidden by Gram reagents. This is evident in the Gram image (Image D) that was taken after reviewing the Gram smear. An oval cell staining pink with a halo around it is clearly seen in the centre of the microscopic field.

The author's experience suggests that such a clue is often useful when numerous cells resembling *Histoplasma* are easily seen in CW, KOH, and histology slides but are difficult to spot in the Gram smear. Sometimes the confirmation comes from experience and not from the specific procedures that are meant to be accurate as a result of greater precision. A Gram stained smear identifying *Histoplasma* was later confirmed by molecular studies done at the reference laboratory.

A few days later, blood was drawn from the patient and sent for fungal culture. The blood specimen was processed and grew white mold after two weeks of incubation on IMA & BHIA. The mold was identified as *Histoplasma capsulatum* after examining the LPAB prep under the microscope, and this was confirmed by DNA probe.

21 CONCLUSION: WHAT YOU WON'T FIND IN THE BOOKS

Medical-laboratory technologists serve as an important instrument between the patient and the clinician. Therefore, it becomes the responsibility of the medical-laboratory technologist to keep thinking about organisms beyond bacteria during the microscopic examination of the direct smears made from the clinical specimens. The medical-laboratory technologists do not have the authority to diagnose patients' clinical conditions or offer consultation to the clinicians; however, they do have the responsibility to provide explicit explanation and interpretation as needed about the microscopic objects seen in the direct smears, such as the Gram stain. The medical-laboratory technologists must take the initiative to at least gather the clinical data and try to understand the relationship between the invading organism and the clinical condition produced by microorganisms, going beyond bacterial entity such as fungi. The Gram stain is the most rapid and cost-effective procedure and is an extremely useful defense mechanism to identify fungi in the direct smears and help the clinicians to promptly treat patients suffering from microbial (fungal) infections.

The Gram stain procedure is commonly used in clinical microbiology for detecting bacteria in clinical specimens. Microbiology staff must be aware of and able to recognize structures other than bacteria present in a Gram smear. Missing organisms in Gram smears would have impact on the patient's diagnosis. The smear reader should have an aptitude to recognize fungal elements (FE) or at least be able to suspect them. All too often, FE remain undetected or missed as microbiology laboratories do not routinely examine direct Gram-stained smears for fungi. Upon observing suspicious-looking objects in the Gram-stained smear, one must ask for an expert opinion or get the Gram smear overstained with CW.

A Gram stain is not an excellent procedure to detect fungi. However, the benefit given upon finding fungi in the Gram stain surpasses any specific procedure suitable for the detection of fungi in the direct clinical specimens. The majority of the Gram smears described in this textbook contained fungi that initially went unreported. One must be aware of the fact that even if medical technologists are not responsible for looking for organisms other than bacteria in the Gram-stained smears, they must keep thinking about "microbiology" during microscopic examination of the Gram smears. In cases when the identity of the structures seen is unclear, it must be brought to the attention of the lab superior or have the specimen sent to a reference laboratory for further work. If a microscopic structure displays a definitive morphology it must not be ignored or discarded but followed through until identified.

A world-famous mycologist confronted the author during a seminar and asked about the use of the Gram stain to find fungi since no medical mycology textbook or journals have ever suggested it. The author asked the mycologist to come and sit in the class and find out what else the Gram stain can do.

Sometimes we prefer a Gram-stained smear that says "NBS." A negative Gram stain report gives us a clue about the presence of an important fungus that does not easily stain by the Gram stain reagents. Hyphal fragments are not regularly picked up in the Gram reagents due to colloidal material masking the FE and preventing the reagents from entering the cell wall, leaving hyphae unstained tracks difficult to spot unless observed carefully.

The author is pleased writing this book for two reasons: First, to clear the doubt about the Gram stain reaction for fungi described as Gram positive in microbiology textbooks. The author has found the irregularity in this statement and is able to hypothesize successfully that the fungi are neither Gram positive nor Gram negative based on the chemistry playing an important role in Gram stain reaction.

There are no medical mycology, clinical microbiology textbooks, or medical journal (except the author's own articles) ever suggested adding Gram smear procedures in a mycology-staining protocol in addition to CW, KOH, GMS, and PAS to detect fungi directly in the clinical specimens. The Gram stain has been proven successful in the institutions where the author has implemented this procedure, and he was the first to publish articles in the United States and Canadian medical journals to press his point. The author's stubborn and strict implementation has changed the diagnosis of several clinical cases where clinical specimens were submitted to the microbiology laboratory requesting C&S only. The author found fungi in the Gram stain smears and helped to change the course of patient's diagnosis and the therapy in a timely fashion. Many cases would have been left undiagnosed if the Gram-stained smears were not specifically checked out for microorganisms beyond bacteria during routine microscopic examination of the clinical specimens.

This book is developed based on experience in the day-to-day running of the medical mycology unit within a microbiology laboratory in an acute-care teaching hospital largely providing health care to the immunocompromised patient population; therefore extra caution is required to search clinical specimens for fungi. Immunocompromised patients are vulnerable to acquire fungal infection that may have severe or fatal consequences if not diagnosed early on and treated appropriately.

The contents of this technical guide are dominated by experience. Experience does not depend on scientific validation. Science itself becomes the self-proclaimed king in the scientific arena. No hypothesis can be developed that would put experience under the test. The scientific techniques remain focused; the end results are likely to disagree at some point. It is for this reason that the science depends heavily on failure. For example, a series of tests is carried out; each time a new thing is discovered, that puts the science in trouble because another scientific proof is required to prove what has already been proven. Experience, on the other hand, does not have such fear, since it is heavily supported by observation. The existence of the experience cannot be formed or discovered by using a scientific test procedure. Readers may be able to read between the lines to find out why Hans Gram and Friedlander continued to disagree with each other on a simple Gram stain. Both were unaware that the organisms they were studying happened to be different. Their experience on two different bacteria (Gram-positive and Gram-negative) led to the useful scientific innovation of dividing bacteria into two distinct categories using the Gram stain procedure.

The reader of this book should be able to conclude why a new edition of the book is necessary and also to know the difference between the knowledge and the wisdom.

22 APPENDIX

22.1 Procedures:

22.2 Gram Stain[1, 2, 3, 4, 7, 39,]

Reagents

Gram stains use basic dyes (colorful cation) such as crystal violet and safranin. Iodine is used to combine with crystal violet (CV) as mordant to lock up crystal violet, making CV-iodine complex a larger-sized molecule unable to be washed away during decolorization. Safranin is the last item in the procedure to stain cells pink that have been rendered colorless due to the use of decolorizer.

Purpose

The purpose of the Gram stain procedure is to detect bacteria in the direct clinical specimen or confirm in culture by categorizing them into two distinct groups: Gram positive (purple) and Gram negative (pink).

Principle

Basic dye such as CV is positively charged and binds to the negative-charged cellular material of bacterial cell, making Gram positive organisms appear as staining purple. The washing step removes excess CV from the smear. Added iodine binds to CV, making the molecule increase in size. Decolorization allows removal of the primary (unbound) stain. Safranin, a counterstain, would bind all types of cells but would show pink color only in Gram negative cells that have been rendered colorless after the use of decolorizer.

Staining procedure

Make a thin smear and air-dry and gently fix with heat. Flood the smear with CV for about 30 seconds. Wash gently with tap water. Flood the smear with iodine for one minute (some suggest 30 seconds). Gently apply decolorizer (acetone/ alcohol) for 10 seconds. Wash momentarily to remove residual from the smear. Flood the smear with safranin for 30 seconds. Wash gently with tap water and leave the smear to dry.

22.3 KOH (10% to 20%)[23, 34, 35]

Purpose:

KOH helps in dissolving keratinized particles and helps in emulsifying solid and viscous material, masking fungal elements present in the specimen. KOH is an unstained and highly sensitive procedure to detect fungal elements present in the clinical specimens.

Method:

- •Place a portion of the specimen in the center of a clean microscopic slide
- •Add two drops of 10–20% KOH and mix it well with the specimen
- •Coverslip the preparation and leave it at room temperature for 15–20 minutes
- •Screen the preparation under the microscope using a low-power x10 objective lens
- •Examine the smear using a high-power x40 objective lens for details if fungal elements are observed during screening

Alternatively: Leave a small portion of the specimen (hard & gritty) in a few drops (0.5-ml) of KOH in a small sterile glass tube. Let it stand for few hours to overnight. The suspension can also be placed in 37°C or higher incubator for one to two hours (or as required) for clearing. Make KOH preparation from the suspension. Examine the preparation under a bright-field microscope using x10 followed by x40 objective lenses.

Interpretation:

Unstained yeast cells with or without pseudohyphae, septate hyphae, and other fungal elements with characteristic morphology. Use flow chart for assistance.

Note: Since KOH is an unstained preparation without contrast, the positive smears must be checked by an experienced technologist, since lot of debris or other material present in the specimen may confuse the smear reader with fungal elements.

22.4 Calcofluor white (CW) by DIFCO-BBL[36]

Principle:

CW reagent contains Cellufluor and is a clear and colorless fluorescent dye that is the disodium salt of 4,4'-bis [4-anilino-6-bis- (2-hydroxyethyl) amino-s-triazin-2-ylamino]-2,2'-stilbenedisulfonic acid. Cellufluor binds beta-linked polysaccharides in chitin and cellulose nonspecifically present in the cell wall of fungal cells. Fungal elements bound with Cellufluor fluoresce when exposed to long-wavelength ultraviolet light. The preparation is read under the UV light microscope equipped with proper exciter and barrier filters (490–520 nm).

Purpose:

The purpose of the CW procedure is similar to the above for KOH preparation, except that it requires the addition of a Calcofluor reagent that binds chitin and cellulose present in the cell wall of fungi.

Method:

•Process specimen as KOH

•Add one or two drops of CW reagent to the specimen mixed with KOH; let stand for a few minutes before reading the smear under the UV-light microscope.

Interpretation:

Fungal cell walls fluoresce bright apple green (refer to Fungi-Fluor stain).

This procedure fails to detect *Pneumocystis jiroveci* (*carinii*).

22.5 KOH vs. CW

- Both nonspecific
- Both highly sensitive
- CW with and without KOH is available and serves as optical enhancer
- CW is a fluorescent stain; optimizes visual clarity
- CW binds surfaces of the cell wall; internal details not usually resolved
- CW would not stain human cells such as macrophages, PMNs
- CW would not display intracellular nature of the organism
- CW may stain cells or objects not really belong to fungal entity but artifacts
- CW use without KOH would miss some organisms
- CW would fail to resolve actinomycetes
- CW with KOH is most useful and beneficial
- KOH is unstained preparation; needs reduced light or phase contrast
- KOH helps dissolving material surrounding FE
- KOH requires experience for reading and interpretation of fungal elements
- High rate of false positive and negative is possible (using KOH) with less-experienced technologists
- KOH provides internal details such as double wall, sporangium, endospores, and nuclear material
- CW and KOH can both be examined under the same microscope equipped with UV and bright light microscope
- Same microscopic field can be visualized as CW and KOH prep by switching the filters in a microscope having UV and white-light systems

22.6 Fungi-Fluor (FF) by Polysciences (same as CW)[5, 23, 36]

Principle:

Cellufluor, the main constituent in the FF reagent "A," is a clear and colorless fluorescent dye that is the disodium salt of 4,4'-bis [4-anilino-6-bis- (2-hydroxyethyl) amino-s-triazin-2-ylamino]-2,2'-stilbenedisulfonic acid. Cellufluor binds beta-linked polysaccharides in chitin and cellulose nonspecifically present in the cell wall of fungal cells. Fungal elements bound with Cellufluor fluoresce when exposed to long-wavelength ultraviolet light. The preparation is read under a UV microscope equipped with proper exciter and barrier filters.

Optional counterstaining regent "B," an aqueous solution of Evans Blue, is provided with the kit to suppress nonspecific fluorescence.

Purpose:

For rapid detection of fungi present in the clinical specimens as well as a rapid and cost effective highly sensitive procedure to detect cyst forms of *Pneumocystis carinii* especially in BAL and induced-sputum specimens (other organs during dissemination)

Method:

- Prepare two thin smears in the center of the microscopic glass slide from specimen
- Allow the smears to air-dry (do not heat-fix the slides)
- Fix smears with absolute methanol for a few minutes until dry. Fixed smears can be stained immediately or stored indefinitely until ready to stain and examine
- Apply two drops (or enough to cover smear area) of FF reagent A for one minute
- Carefully spread the reagent evenly over the specimen area with a wooden stick
- Rinse slide with tap water (or de-ionized water)
- Apply a coverslip on the smear while wet and examine smear under the UV light microscope. Alternatively, store stained slides in the dark and re-hydrate dry smear with a drop of distilled water and coverslip and examine the smear as above.
- Apply a few drops of optional counterstain solution B for thicker smears and quickly rinse with tap water. Coverslip the preparation and read as above.
- Use low-power x25 for screening and high-dry x40 objective lenses for observing details

Quality Control:

Add a known *Candida albicans* smear with each lot when running CW and FF.

Optional: Occasionally use a known smear containing septate or non-septate hyphae.

22.7 Interpreting fungal elements seen under the microscope using FF

Intense peripheral staining and characteristic morphology helps identifying fungal elements. Fungal cell walls fluoresce bright apple green. Observe structures and define their shape and size. Refer to the flowcharts 1 and 2 (section 2.2) to help interpreting fungal elements appropriately. Counterstain used would make the background red orange, and the fungi would fluoresce yellow green.

Pneumocystis jiroveci (*carinii*) will appear as 5–7 µm spherical bodies with intensely staining bean-shaped or double-parenthesis-like internal structures with opposite sides flattened, surrounded by a faint cyst wall.

NB: Gram-stained smears can be overstained by CW or FF after removing oil from slide using xylene and alcohol. Similarly, FF-stained slides can be overstained with specialized stains such as GMS, PAS, Giemsa, mucicarmine, and Fontana Masson.

Other objects that can fluoresce with CW and FF are collagen, elastin, cotton fibers, plant material, some cells, cell inclusions, and parasites (e.g., *Acanthamoeba*).

Pneumocystis jiroveci (*carinii*) cysts are detected by FF (Polysciences).

Other routine simple staining procedures such as Gram stain, ZN, and MK are done in routine microbiology. Most fungi when present can be detected in Gram-stained smears. Its usefulness has a great impact on patient care when fungi are detected in the Gram-stained smears but the clinicians made no request to investigate fungi in the patient's clinical specimens. Technologists reading smears looking for bacteria must be aware of the presence of fungi in the Gram-stained, ZN and MK smears. An India Ink test is usually done on CSF and respiratory specimens such as BAL and also done on Cryptococcal cultures.

Specialized stains such as GMS, PAS, H&E, mucicarmine, and Fontana Masson are not carried out routinely in mycology laboratory. Such procedures are done on demand by sending the smear to histopathology for specific staining procedures.

22.8 Procedure for overstaining Gram with KOH/CW (author's own procedure)[31]

- Gently wipe off oil from Gram-stained smear using a clean tissue paper
- Apply xylene to the Gram smear covering staining area for one to two hours
- Rinse the slide with absolute methanol
- Leave methanol on the smear for thirty minutes or longer if required
- Rinse smear with 70% alcohol and leave it on for thirty minutes
- Rinse smear with distilled water and leave it on for thirty minutes
- Rinse smear with 10 or 20% KOH and leave it on for thirty minutes or longer if required
- Pour off KOH, add a drop of CW reagent, and mix gently with wooden applicator
- Apply coverslip, gently press it down, and examine the preparation under the UV microscope equipped with proper filter system
- Observe that "fungal elements" that have previously been suppressed by Gram reagents would now stand out, fluorescing with various shades of apple green depending on the amount of the reagent reaching the fungal cell wall
- Read preparation as KOH under a bright-field microscope and under a UV light microscope for CW

23 References

1. Adams, E. "Studies in Gram staining." *Stain Technology.* 1975;50(4):227–231.

2. Atkins, K. N. "Report of committee on descriptive chart. Part III. A modification of the Gram stain." *J. Bacteriol* 1920;5:321–324.

3. Bartholomew, J. W.; Roberts, M. A.; Evans, "E. E. Dye exchange in bacterial cells and the theory of staining." *Stain Technology.* 1950;25(4):181–186.

4. Bartholomew, J. W. "Variables influencing results, and the precise definition of steps in Gram staining as a means of standardizing the results obtained."

Stain Technology 1962;37(3):139–155.

5. Baselski, V. S., et al. "Rapid Detection of *Pneumocystis carinii* in Bronchoalveolar Lavage Samples by Using Cellufluor Staining." *Journal of Clinical Microbiology.* 1990, February;28:393–394.

6. Boggild, A. K.; Poutanen, S. M.; Mohan, S.; Ostrowski, M. A. "Disseminated phaeohyphomycosis due to *Ochroconis gallopavum* in the setting of advanced HIV infection." *Medical Mycology.* 2006, December;44:777–782.

7. Bryskier, A., ed. *Antimicrobial Agents.* Washington, D.C.: ASM Press, 2005.

8. Chandler, F. W., and Watts, J. C. *Pathologic Diagnosis of Fungal Infections.* Chicago: ASCP Press. 1987.

9. Collier, L., Balows, A., Sussman M. *Topley & Wilson's Microbial Infections, Ninth Edition,* Volume 4, Medical Mycology. Arnold, a member of the Hodder Headline group, Great Britain, 1998.

10. De Hoog, G. S.; Guarro, J.; Gene, J.; Figueras, M. J. *Atlas of Clinical Fungi, Second Edition.* Utrecht, the Netherlands: The Centraalbureau voor Schimmelcultures (CBS), 2000.

11. Dickey, N., ed. *Funk & Wagnalls New Encyclopedia.* Volume 9. New York: Funk & Wagnalls Corporation, 1986, p. 122.

12. Edberg, S. C., et al. "Esculin-Based Medium for Isolation and Identification of *Cryptococcus neoformans.*" *J. Clin. Microbiol.* 1980;12:332–335.

13. Forbes, B. A.; Sahm, D. F.; Weissfeld, A. S. Bailey and Scott. *Diagnostic Microbiology. Tenth Edition.* Mosby, Inc. USA, 1998.

14. Forbes, B. A.; Sahm, D. F.; Weissfeld, A. S. *Bailey & Scott's Diagnostic Microbiology, Eleventh Edition.* St. Louis, MO: Mosby, 2002.

15. Girmenia, et al. "*Candida guilliermondii* Fungemia in Patients with Hematologic Malignancies." *J. Clin. Microbiol.* 2006;44:2458

16. Hucker, G. J. "A new modification and application of the Gram stain." *J. Bacteriol.* 1921;6:395–397.

17. Jacobson. "W. Gram's discovery of his staining technique." *J Infect.* 1983;7(2):97–101.

18. Jacobson. "W. Historical View on Gram." *J. Infect.* 1983;7, 97–101.

Twenty-Ninth Annual Meeting of the Society of American Bacteriologists. School of Medicine and Dentistry, University of Rochester, Rochester, New York. December 28, 29, and 30, 1927.

19. Kini U. and Babu, M.K. "Ocular spherulocystosis." *Journal of Clinical Pathology.* 1996;49:857–858.

20. Koneman, E. W.; Allen, S. D.; Janda, W.M.; et al. *Color Atlas and Diagnostic Microbiology.* Philadelphia, PA: Lippincott, 1997, pp. 11–14, 102.

21. Kopeloff, N. and Cohen, P. "Further studies on a modification of Gram stain." *Journal of Bacteriology.* 1928;15:12.

22. Kumar, D.; Sigler, L.; Gibas, C.; Fe, C.; Mohan, S.; et al. "*Graphium basitruncatum* Fungemia in a Patient with Acute Leukemia." *J. Clin. Microbiol.* 2007, May;45(5):1644–1647.

23. Larone, D. H. *Medically Important Fungi. A Guide to Identification, Fourth Edition.* Washington, D.C.: ASM Press, 2002.

24. Lee, S. B.; Oliver, K. M.; Yi, N. J.; Strube, M. S.; Mohan, S. K.; Slomovic, A. R. "Fourth-generation fluoroquinolones in the treatment of mycobacterial infectious keratitis after laser-assisted in situ keratomileusis surgery." *Canadian Journal of Ophthalmology* 2005;40(6).

25. McClatchie, S.; Warambo, M. W.; Bremner, A. D. "Myospherulosis: a previously unreported disease?" *American Journal of Clinical Pathology* 1970;51:699–704

26. *Medical Mycology.* International Society for Human and Animal Mycology 2006;44, supplement 1. Taylor & Francis Group.

27. *Medical Mycology.* International Society for Human and Animal Mycology 2005;43, supplement 1. Taylor & Francis Group.

28. Mills, S. E. and Lininger, J. R. "Intracranial myospherulosis." *Human Pathology*-Elsevier 1982;13:596–597.

29. Mims, C.; Playfair, J.; Roitt, I.; et al. *Medical Microbiology, Second Edition.* London: Mosby International, 1998, pp. 25–27.

30. Mohan, S. K., et al. "A case study of *Coccidioides* immitis: New approaches to identifying an old bug." *CJMLS.* 2007;69:98–112.

31. Mohan, S. K. "Beyond Bacteria: Interpreting Fungal Elements in the Gram Stain." *CMNJ.* 2004;26(14):108–112.

32. Mohan, S. K. "Incidental detection of parasites in a mycology laboratory during routine microscopic examination of smears stained by Fungi-Fluor." *CJMLS.* 1996;58:151–155.

33. Murray, P. R.; Baron, E. J.; Jorgensen, J. H.; Landry, M. L.; Pfaller, M. A. *Manual of Clinical Microbiology, Ninth Edition.* Volume 2. Washington, D.C.: ASM Press 2007.

34. Murray, P. R.; Baron, E. J.; Pfaller, M. A.; Tenover, F. C.; Yolken R. H. Manual of Clinical Microbiology, Seventh Edition. Washington D.C.: ASM Press, 1999, pp. 1346, 1675.

35. Murray, P. R., et al. *Manual of Clinical Microbiology, Sixth Edition.* Washington D.C.: ASM Press, 1995.

36. Perry, J. L. and Miller, G. R. "Quality control slide for potassium hydroxide and cellufluor fungal preparations." *J. Clin. Microbiol.* 1989, June;27(6): 1411–1412.

37. Pfaller, et al. "*Candida rugosa*, an Emerging Fungal Pathogen with Resistance to Azoles." *J. Clin.*

Microbiol. 2006;44:3578

38. Pfaller, et al. "*Candida guilliermondii*, an Opportunistic Fungal Pathogen with Decreased Susceptibility to Fluconazole." *J. Clin. Microbiol.* 2006;44:3551

39. Popescu, A. and Doyle, R. J. "The Gram Stain after more than a century." *Biotechnic and Histochemistry.* 1996;71(3).

40. Rosai, J. "The nature of myospherulosis of the upper respiratory tract." *American Journal of Clinical Pathology* 1978;69:475–481.

41. Saksun, J. M.; Kane, J.; Schaacter, R. K. "Mycetoma caused by *Nocardia madurae.*" *Can Med Assoc J.* 1978;119:911–914.

42. St. Germain, G. and Summerbell, R. "Identifying Filamentous Fungi." *A Clinical Laboratory Handbook.* Star Publishing Company, Belmont, California, USA. 1996.

43. Tang, P.; Mohan, S.; Sigler, L.; Witterick, I.; Summerbell, R.; Campbell, I.; Mazzulli, T. "Allergic Fungal Sinusitis Associated with *Trichoderma longibrachiatum.*" *J. Clin. Microbiol.* 2003, November;41(11):5333–5336.

44. Travis, W. D.; Li, C. Y.; Wieland, L. H. "Immunostaining for hemoglobin in two cases of myospherulosis." *Arch. Pathol. Lab. Med.* 1986;110:763–765.
45. Webster, J. *Introduction to Fungi, Second Edition.* Cambridge, Great Britain, Cambridge University Press, 1980.

46. Welsh, O. "Mycetoma." Article Last Updated: Feb 22, 2007 http://www.emedicine.com/derm/topic280.htm

47. Wheeler, T. M.; Sessions, R.; McGavran, M. H. "Myospherulosis: A preventable iatrogenic nasal and paranasal entity." *Arch. Otolaryngol.* 1980;106:272–274.

Printed in the United States
By Bookmasters